LANDFORMS of IOWA

A Bur Oak Original

LANDFORMS of IOWA

by

Jean C. Prior

designed and illustrated by

Patricia J. Lohmann

University of Iowa Press
Iowa City

for the
Iowa Department of Natural Resources

University of Iowa Press, Iowa City 52242
Printed in Hong Kong
First edition, 1991

Printed on acid-free paper

Library of Congress
Cataloging-in-Publication Data
Prior, Jean Cutler.
Landforms of Iowa/by Jean C. Prior.
p.   cm.—(A Bur oak original)
Includes bibliographical references and index.
ISBN 0-87745-350-0 (acid-free paper),
ISBN 0-87745-347-0 (pbk.)
1. Geomorphology—Iowa.   I. Lohmann,
Patricia J.   II. Iowa. Dept. of Natural
Resources.   III. Title.   IV. Series.
GB428.I6P73   1991   91-16136
551.4′1′09777—dc20   CIP

Acknowledgments

Sincere appreciation is expressed to Donald L. Koch, state geologist and chief of the Geological Survey Bureau; to Larry L. Bean, administrator of the Energy and Geological Resources Division; and to Larry J. Wilson, director of the Iowa Department of Natural Resources, all of whom supported this edition and coordinated the arrangements for its publication. The exceptional images within are the generous work of the photographers noted, with special acknowledgment to Gary Hightshoe, Iowa State University; Drake Hokanson, formerly with the University of Iowa; Donald Poggensee of Ida Grove; Douglas Harr, Iowa Department of Natural Resources; and the "Land between Two Rivers" group at Iowa Public Television. The University of Iowa Dental Media group applied their skills to ice-age rodent jaws and mammoth molars. The high-quality illustrations and overall visual design crafted by artist Patricia J. Lohmann speak for themselves; her sense of style and insights into communication of information to the public are recognized and appreciated. Thanks are expressed to my colleagues Brian J. Witzke, for his expertise in large Pleistocene fauna, and Bernard E. Hoyer, Timothy J. Kemmis, and E. Arthur Bettis III for their long-standing support of this edition, numerous consultations, and detailed reviews of the manuscript.

Finally, I am grateful for the opportunity to prepare this book. The public and scientific community's response to *A Regional Guide to Iowa Landforms* in 1976 and the period of productive Quaternary research in Iowa since then fueled the decision to allocate time to preparing another volume. The opportunity to study the Earth, to inform and perhaps influence readers, while encouraged by talented, good-natured colleagues is a rare and appreciated combination.

Contents

Maps of Iowa

Preface

The study of geology shows us another world just beneath the familiar scenes of our everyday lives. It draws us outdoors to interesting places as we probe the depths of Earth history and the great antiquity of geologic time. The landscapes we observe are three-dimensional; their breadths, elevations, and depths are the culmination of a fascinating array of geologic events and processes. A knowledge of these processes is important to a good education, to responsible land and water stewardship, and to the quality of our lives. The purposes of this book are to awaken an interest in Iowa's landscapes and to develop an awareness of their diversity. The book is intended to provide a basic reference for Iowans and visitors to the state who seek explanations for Iowa's landscape features and the geologic ingredients beneath them.

In the last fifteen years, we have made significant progress in understanding the geologic contents of Iowa landforms and the geological processes that have stirred the state's land surface. Drilling, dating, and mapping together with stratigraphic, paleontologic, and geomorphic analyses have established a more complete, more coherent view of the geologic evolution of the Iowa landscape. To prepare this account, I have mined the recent technical literature, especially the contributions of George R. Hallberg, Timothy J. Kemmis, E. Arthur Bettis III, Bernard E. Hoyer, Richard G. Baker, Holmes A. Semken, Jr., R. Sanders Rhodes II, Donald P. Schwert, Terrence J. Frest, and John D. Boellstorff.

The year 1992 marks one hundred years of continuous, state-supported geological investigations in Iowa. It is important to communicate the results of these investigations to the public and to those working with Iowa's natural resources. This book is a view of what we know at this point, not the final word on the subject. Future studies and discoveries will unearth new sites and fresh insights that will modify and clarify our knowledge of the geological basis of Iowa's land resource.

I once heard natural-history author John Madson comment that to take the time to show others something of interest in their surroundings is to give a gift for life. This book represents my endorsement of that philosophy. I dedicate it to the people of Iowa.

LANDFORMS of IOWA

Shapes and features of the land
surface vary from one region of Iowa
to another. Ridged summits alternate
with steeply pitched, prairie-covered
slopes west of Smithland in
Woodbury County. This sharp-
featured terrain is sculptured from
thick deposits of windblown silt swept
from the Missouri River valley during
glacial episodes.

Photo by Donald Poggensee

Introduction

Any single view of the Iowa landscape takes in a distinct collection of landforms. The recognizable shapes and features of the land's surface change from one part of the state to another, and Iowa can be divided into regions based on these associations of individual landforms. What are the state's characteristic landforms and what do they look like? What landscape features identify a particular region of the state and distinguish it from other regions? What are these landforms made of, and how and when were they formed? Such questions about appearances and origins arise naturally in the mind of any curious observer.

The shapes seen in the Iowa landscape are inherited from the geologic past. Just as a person's face reflects something of the tranquillity and turmoil of past years and events in life, so too do the creases of the Iowa landscape reflect a fascinating history of geologic events. The geologic forces that shaped the state's land have left their signatures within the deposits beneath the ground and among the contours of the landscape itself.

These geologic settings influence the distribution of native plant and animal communities in Iowa. They also determine an area's vulnerability to contamination problems or natural hazards to human activity. Iowa's land supports the construction of buildings, the planting of crops, the burial of wastes, the pumping of groundwater supplies, and the extraction of mineral resources. In addition, we need and enjoy the personal refreshment of meeting the land's beauty on its own undisturbed turf. It is important, therefore, to understand the land and how it functions, now and through time.

The following chapters will introduce how people have examined and recorded the Iowa landscape throughout history. I will delve into those geological events having the most significant impact on the physical appearance of the state and consider each of Iowa's distinctive landform regions. Finally I will identify places to visit in order to see notable local effects of the geological processes that molded and stamped the state's land surface.

Outlooks
on Iowa Landforms

Then as to scenery (giving my own thought and feeling), while I know the standard claim is that Yosemite, Niagara Falls, and the upper Yellowstone and the like, afford the greatest natural shows, I am not sure but the Prairies and Plains, while less stunning at first sight, last longer, fill the aesthetic sense fuller, precede all the rest, and make North America's characteristic landscape.

—Walt Whitman
"Specimen Days," 1881

Landmark—the very word echoes a fundamental relationship between humanity and distinctive features of the terrain. For generations people have looked to the landscape to guide their travels, inspire their songs, provide their living, diversify their recreation, and define their geographic home.

Iowa's land is not marked by rugged mountain peaks, steep canyons, or torrential waterfalls. An eye accustomed to dramatic scenery can rest in Iowa and will gradually become fine-tuned to less intense topographic irregularities. Out of these gentler landscape patterns, however, differences are noticed and, indeed, Iowa's own distinct landmarks take shape—the bend of a river, the slope of a hillside, a notch along the horizon, a sentinel limestone bluff, the flow from a spring, a knob of gravel, an isolated boulder, a cluster of marshes, or a band of silty ridges. This land, including the geologic deposits layered beneath its surface, has touched the lives of residents and visitors, past as well as present.

The examination of Iowa's terrain reveals a fascinating chronicle of people and their endeavors. Shifting trends in the nation's history, developing concepts in the field of geology, and Iowa's own geologic setting and resources have influenced the reasons for and methods of looking closely at Iowa's landscapes.

Prehistoric Indian tribes who once lived on this land knapped tools from flint found in bedrock ledges. They buried their dead on high ground overlooking river valleys and in earth which they shaped into mounds. Later, as the frontier of the United States beckoned European explorers westward, Iowa's bordering rivers became important avenues into the continent's interior. Marquette and Jolliet (1673), Julien Dubuque (1788), Zebulon Pike (1805), Thomas Nuttall (1809), Henry Schoolcraft (1820), George Catlin (1835), and Jean N. Nicolet (1838) made their way into the Mississippi Valley. Lewis and Clark (1804), Nuttall (1810), Prince Maximilian and Karl Bodmer (1833), and Catlin (1836) traveled the waters of the Missouri. These men entered their observations about the land into journals or drew them in

Report on the Geological Survey of the
State of Iowa, Vol. 1, 1870

*In the nineteenth century, geologists relied on their artistic skills to record landscape features and
geologic strata. This 1868 illustration by Orestes St. John shows outcrops of Mississippian limestone
along the picturesque valley of Flint Creek, now the site of Starrs Cave State Preserve in
Des Moines County.*

sketchbooks, and these historic records became the earliest references to Iowa landforms.

This era of Euro-American discovery and exploration was followed by settlement and more purposeful geological study. The Iowa Territory was established in 1838, and shortly after the United States government commissioned David Dale Owen to collect information on mineral resources of the lands drained by the upper Mississippi River. His results were published in 1852 under the title *Report of a Geological Survey of Wisconsin, Iowa and Minnesota*, which is regarded as the first official geologic investigation in Iowa. Owen's 639-page monograph is richly illustrated with stylized sketches of landscapes, detailed drawings of fossils, and colored cross-sectional profiles of river valleys including the Mississippi and Des Moines. Among his considerable talents, Owen was a skilled artist, and like many of the geologists of this period, he received his science education as part of his study of medicine.

Into the land between the Mississippi and Missouri rivers journeyed Iowa's pioneer settlers, who viewed the vast open-rolling prairie with mixed feelings, depending on whether the home forests of the eastern states were regarded as confining or sheltering. Passage through the open land was slow and difficult, as these early travelers were confronted with the hazards of marshlands, steep hills, river crossings, and rock-strewn fields. The very slowness of travel, however, prompted their observation of the details in the surrounding landscape. A prominent ridge on the horizon, the place where rivers joined, the shape of an unusually large boulder, and the color of local rock or soil became landmarks to guide their progress across the state. Those who stayed found fertile soils beneath the tallgrass prairie cover.

In 1855 legislation was passed providing for the first state-supported geological survey of Iowa. The fledgling state geological surveys in the Midwest were motivated by the search for natural resources and the need for rapid reconnaissance of the land. Geologists associated with the first Iowa survey were men who later went on to make national contributions during their careers—James Hall, Josiah D. Whitney, Amos H. Worthen, and Fielding B. Meek.

There was a closeness between science and art during this part of the nineteenth century. The tradition of topographical landscape artists in America included not only professional artists but those within the scientific community who drew landscapes as a means of conveying geological information to the readers of their reports. The grandeur of the mountain ranges in the western United States was first seen through the panoramic drawings and paintings of the documentary artists who accompanied the geological expeditions of Ferdinand V. Hayden, Clarence King, John Wesley Powell, George Wheeler, Clarence Dutton, and Grove Karl Gilbert. Thomas Moran and William Henry Holmes became well known for their impressive landscape artistry and William H. Jackson for his pioneering photographic work while accompanying these important surveys of the American West.

Orestes St. John was a geologist for the state of Iowa between 1866 and 1869 who later worked with Hayden

Black and white photographs of Iowa's landscape illustrated the state's geological reports beginning in 1892. This photograph by Samuel Calvin, in the 1898 "Geology of Delaware County," shows the effects of weathering on creviced Silurian dolomite. Today, these striking outcrops are part of Backbone State Park.

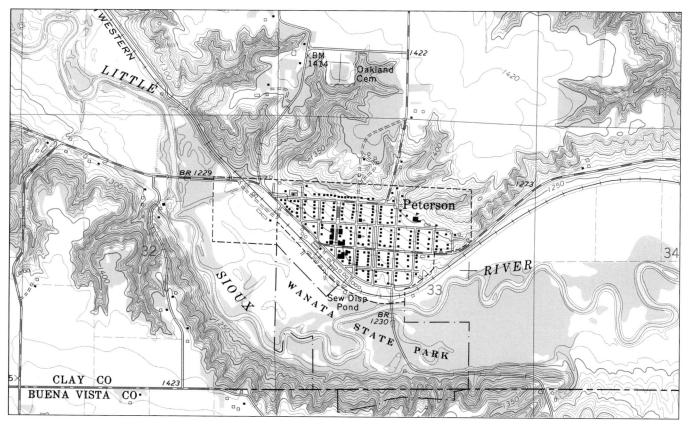

U.S. Geological Survey, Peterson Quadrangle, 7.5 Minute Series, 1971

Topographic maps are accurate, detailed portrayals of the land surface. Variations in elevation are shown by brown contour lines; water is blue, woodlands are green, and highways and cultural features are red or black.

in New Mexico, Idaho, Wyoming, and Colorado. Like Owen, he relied on his skills as an artist to document geologic features and resources. His sketches and lithographs have historical interest, artistic validity, and scientific value as early illustrations of Iowa landforms (lithograph, p. 3).

Early views of Iowa's land prior to the widespread availability of cameras were portrayed by geologists who were motivated by the natural features they observed, by what various contours, clays, boulders, fossils, and rocks revealed about the geologic processes that shaped the land, and by the economic resources associated with those features and deposits. These early observers were naturalists in the classic sense of the word. Their pioneering descriptions of the land and theories on its geological evolution were but part of a wide-ranging interest in the Earth's surface. They spent long days in the field traveling on foot and horseback. This close contact with the land resulted in significant observations of plants, animals, and prehistoric cultures in addition to geological fact-finding.

A permanent Geological Survey was established as an agency of Iowa's state government in 1892, and many county geological reports published in its *Annual Report* series during the late 1800s and early 1900s were supplemented with extensive botanical notes on prairie and forest flora as well as meteorological records or information on archaeological remains. In fact, the *Bulletin* series, published between 1901 and 1930, devoted entire volumes to the grasses, weed flora, rodents, raptorial birds, and honey plants of Iowa.

These geological reports were authored by noted scientists like Samuel Calvin, Thomas Macbride, and Bohumil Shimek who were equally at home in several fields of natural history now regarded as separate scientific disciplines. The engaging geological accounts by these men and other early Iowa geologists such as Charles R. Keyes, William H. Norton, and H. Foster Bain were written in a personal, almost poetic style seldom seen in today's technical literature. For example, in "Geology of Harrison and Monona Counties," Shimek described the bluffs and ridges bordering the Missouri River valley:

> During the day these bluffs may burn in the heat of the midday sun, they may be swept by the hot blasts of summer winds, or hidden in the whirling clouds of yellow dust which are carried up from the bars of the great river; but in the stillness of early morning, and again when the peace and quiet which portend the close of day have settled upon them, they are both restful and inspiring when looked upon from the valley; and there is no grander view than that which is presented under such circumstances from their summits,—on the one hand over the broad valley and on the other across the billowy expanse of the inland loess ridges which appear like the giant swell of a stormy sea which has been suddenly fixed.

In addition, these volumes were illustrated by some of the earliest photographs taken of the Iowa landscape (photo, p. 5). Approximately 7,000 glass plate negatives and prints from the period 1875 to 1925 are preserved in the Calvin Collection at the geology department of the University of Iowa.

Following the era of the geologist-naturalist, the

Aerial photography provides views of the landscape from the perspective of altitude. This color-infrared photo of the meandering Big Sioux River, its floodplain, and adjacent Loess Hills was taken in April 1980 from 40,000 feet, about 17 miles northwest of Sioux City.

Iowa Department of Natural Resources, Geological Survey Bureau

study of the shapes of the Earth's surface, known as geomorphology, became a more specialized field within the professions of geology and geography. With the evolution from naturalist to specialist and with improvements in travel and field techniques came a shift from line drawings and cumbersome cameras to topographic maps and aerial photographs as important methods to illustrate and examine Iowa landforms.

Topographic mapping of Iowa began in the 1880s, but until 1970 only limited areas of the state were covered on maps of various scales. In that year the U.S. Geological Survey, in cooperation with the Iowa Geological Survey, began a concentrated effort to finish map coverage of Iowa at the 1:24,000 scale—the popular 7.5 minute quadrangle. This program was completed in 1986 and provides the most accurate and detailed maps of Iowa's land surface available today.

Topographic maps are useful because they provide the viewer with a means of observing the land surface in three dimensions (map, p. 6). Variations in the land surface are illustrated by the use of contour lines, which connect points of equal elevation along the land surface. With a little practice the viewer can use the arrangement and spacing of contour lines to get a clear picture of the terrain with its hills, valleys, and drainageways; one can easily visualize the relief, or inequalities of the land surface, and the shapes of individual landforms. These maps are invaluable for planning outdoor recreation activities or assessing the terrain in an unfamiliar area, for classroom orienteering activities, and as a basic tool in a number of field-oriented sciences.

With the advent of aircraft, a whole new outlook on Iowa's land became possible. In contrast to the graphic representation of the land surface on topographic maps, aerial photographs are records of actual views of the terrain. Shadows, soil moisture, vegetation, drainage patterns, and land use are features of aerial photos that can be used to advantage to study landforms and the materials that compose them. Many rural residents are familiar with the 9 by 9-inch black-and-white prints available from the U.S. Agricultural Stabilization and Conservation Service (ASCS) and the U.S. Department of Agriculture Soil Conservation Service (SCS). A combination of aerial photography and soil mapping is found in the county soil surveys produced by the SCS. These highly detailed maps of soil types are printed on 1:15,840-scale aerial photographs and are an invaluable tool for landform and surficial geology studies in Iowa. In addition, high-altitude (40,000 feet) color-infrared aerial photographs of Iowa provide exceptional views of the landscape and land-use patterns (photo, left).

Recorders of Iowa landforms have ranged from artists to astronauts. Today numerous practical benefits of space exploration result as cameras in space focus back on Earth. Images from orbiting satellites and spacecraft provide time-lapse views of extensive areas of the Earth's surface. The Iowa landscape can be seen from a perspective never before available, through the "eyes" of different camera-film combinations or other sensors. At distances ranging from 100 to 500 miles above the Earth, regional terrain patterns stand out, enhanced by drainage networks, soil types, vegetative cover, and land use (satellite image, right). This distant examination of the Earth's surface, called remote sensing, is used not only by those studying the shapes and patterns of terrain features but also by those interested in mineral resources, water quality, forestry, wildlife management, and urban and transportation planning.

Looking at landscapes from the personal view of a field sketch or from the remoteness of space tells us much about surface forms and patterns of distribution. To obtain a more thorough understanding of landforms and their origins, however, we must also look inside them to see what types of materials lie beneath the ground, giving shape and durability to its surface. The interiors of landscapes can be examined wherever streams erode into hillsides or where earth materials are exposed in roadcuts or quarries. In general, such exposures in Iowa are few and far between. Our knowledge about the internal aspects of the state's landforms often comes from drilling wells, especially for groundwater supplies. Rock and soil samples recovered from drill sites are used to compile three-dimensional pictures of the variety, thickness, and distribution of deposits that underlie the landscape.

There is still one more important perspective from which landforms are viewed—that of time. Iowa's landscapes have varied dramatically in their appearance from one period of geologic time to another. Consider, for example, that Iowa was once part of a vast inland sea containing great numbers of marine organisms now found as fossils in the rock record. Millions of years later the state was locked in a deep-freeze by hundreds, perhaps thousands of feet of glacial ice. Looking at Iowa today,

note how we tend to think of the landscape as being permanent and unchanging. The strikingly different environments that existed in the state's geologic past were the result of small shifts and adjustments in enduring Earth processes. We notice the dramatic, rare occurrence of a catastrophic flood or a sudden rock slide, but with our nearsighted view of time we cannot casually sense those weathering and erosional processes that continue steadily day after day, year after year, gradually changing the face of the land in the course of thousands, or tens of thousands, or millions of years. Even those infrequent, spectacular events that take us by surprise, producing massive alterations of the landscape in brief periods of time, actually occur with geologic regularity.

Geologic time is not clocked by intervals such as minutes, hours, or days. We may think of time as measured in these units of equal duration, but our personal view of time is more likely tied to events such as birthdays, graduations, weddings, or vacations—events of unequal spacing to which we relate other, less important happenings. The passage of geologic time is also marked by events of importance such as volcanic eruptions, trespassings of ancient seas, episodes of mountain building, evolution and extinction of prehistoric species, or the ex-

Regional differences in terrain and drainage patterns across Iowa are revealed in this computer-enhanced satellite view. The image was taken in May 1984 from 496 miles above the Earth using a sensor responsive to exposed soils (reddish brown to gold) and permanent vegetation (shades of green).

0   20   40   60 mi.

0   40   80 km.

Prepared by
James Giglierano from a
National Oceanographic and
Atmospheric Administration
image

Satellite Image of Iowa

pansion and melting of continental ice sheets. These events form a set of reference points from which a story can be told.

One of the great gifts of geology to general thought is the perspective it provides on the universe, on the Earth, and on life. An appreciation for the concept of geologic time makes us think more carefully about the land. It allows us to consider some of the broader patterns at work that are concealed by viewing the future only one year at a time or by looking at the past just in terms of a single lifetime. The attention focused today on groundwater quality, soil conservation, energy supplies, waste disposal, urban growth, and protection of native habitats reminds us of the vital role that Iowa's land, water, and mineral resources play in the quality of our lives. We must realize there are limits to these resources. They are not uniformly distributed in quantity or quality, and there is competition for their use. The resolution of the environmental issues of our time and the protection and management of the state's natural resources hinge on understanding and appreciating the geologic systems that operate on and beneath the land.

The mounds of prehistoric Indians, the prose and poetry of writers, the sketches and paintings of landscape artists, the interpretations of earth scientists, and the seasonal harvests by generations of farmers are the inspirations of diverse outlooks toward the land. In turn, these people have passed along something of the reverence, mood, character, history, and productivity gleaned from their contacts with Iowa's land (photo, left).

A person's outlook on the landscape can inspire a wide range of thoughts and activities. Here, the photographer's artistic sensitivity to composition, color, patterns, light, and shadows resulted in this striking scene of a Johnson County field.

Photo by Drake Hokanson

13

Geologic Origins
of Iowa Landforms

The historic interest of an ancient document does not depend on the size of its letters, nor are the geological values of the landscapes . . . lessened by the faintness of the characters in which their story is engraved.

—William H. Norton
"Geology of Cedar County," 1901

All landscapes speak about their geologic pasts. While mountain peaks and canyon depths boldly hail us with rugged, colorful proclamations of their origins, Iowa's land quietly narrates its geological legends for those who stop to listen. The state has a geologic past of immense duration that includes spectacularly different environments which previously occupied the same geographic area. That past is packaged in layers within the Earth, with the latest events still visible in the shapes and deposits exposed at the land surface. These landscapes and materials record the remains of ancient tropical sea floors, the passage of glaciers, the accumulation of wind-blown sediment, and the sculpture of flowing water. To look at Iowa's landscapes with a greater understanding of their geological roots, it is useful to step back and examine the full range of geologic events and processes that have influenced the state's terrain.

The answers to many questions about landform shapes and landscape origins are tied to the types of materials found beneath their surfaces. It is always intriguing to pick up a rock, a fossil, or some loose dirt and wonder about its age. Many earth materials and fossil remains contain built-in clocks which geologists can use to measure geologic time. Such geologic timekeepers are possible because certain radioactive elements in nature have their own long-term but predictable rates of breakdown into more stable elements or isotopes. As time passes, the increasing number of by-products of this breakdown can be measured against the remaining number of parent elements, and the amount of time elapsed can be determined. Different element pairs, especially rubidium-strontium, potassium-argon, uranium-lead, uranium-thorium, and carbon-nitrogen, can be selected to measure the age of different earth materials, ranging from the most ancient rocks to very young deposits, including even the charcoal of prehistoric Indian encampments. These radiometric dating techniques give geologists the precision to tie the Earth's deposits and related geologic events to actual increments of time (photo, right).

Consider for a moment that the Earth has been evolving for about 4.5 billion years, as determined by dating of meteorite fragments and lunar rock samples.

The passage of geologic time can be measured using chemical isotopes present in cave deposits. Stalactites in Winneshiek County's Cold Water Cave began to grow about 160,000 years ago.

The most ancient rocks lying in Iowa's geological basement are igneous and metamorphic varieties composed of interlocking mineral crystals. These Precambrian-age rocks originated from molten materials that oozed from the Earth's interior and then cooled and often were contorted by the pressures of continent-building activity. Measurements of radioactive isotopes locked in mineral crystals taken from deep cores of these crustal rocks in Iowa have yielded dates as old as 1.4 billion years in Jackson and Cherokee counties, 1.7 billion years in Sioux County, and—some of the continent's oldest rocks—2.5 billion years from a core drilled in southeastern Lyon County, north of Matlock.

Among these deeply buried basement rocks, a striking feature of the ancient Iowa landscape has been recognized—the Midcontinent Rift System. This geologic tear in the Earth's crust extends from Lake Superior south and west into central Kansas; its existence across Iowa, from Worth County to Mills and Pottawattamie counties, is known from measuring the variations in gravitational pull and the differences in magnetic intensity across the area. Dense volcanic rocks that erupted into the opening along this seam were dated at 1.02 billion years in Guthrie County. Iowa's deepest well, drilled in Carroll County in 1987, was an exploration test for petroleum in the ancient sedimentary strata that flank the rift. Zircon crystals in the igneous rocks encountered at the bottom of this 17,851-foot well yielded a date of 1.28 billion years.

Iowa's Precambrian rocks rise unusually close to the land surface at two different locations. In the vicinity of Manson, near the Calhoun-Pocahontas county line (map, right), an area of igneous and metamorphic rocks was lifted to the surface by the explosive shock of a meteorite impact about 66 million years ago. The 22-mile-wide circular crater is not visible at the land surface because it is covered with as much as 300 feet of glacial deposits. Its presence, however, is known from wells that tap the naturally soft groundwater in the fractured Precambrian granite. In contrast, almost no groundwater supplies are found in the chaotic mixtures of deformed and faulted sedimentary rocks that surround the granite core. The second location is in the northwest corner of Lyon County at Gitchie Manitou State Preserve. Here, in a rare appearance, basement rocks actually poke through to the land surface. These unusual Iowa outcrops, composed of the distinctive reddish Sioux Quartzite, were dated between 1.6 and 1.7 billion years old.

After the Precambrian foundation became established, North America huddled near the equator, close to most of the planet's other major land masses. Over the next 600 million years much of the continent was periodically submerged beneath warm, shallow seas that flooded inland from the open oceans. The sedimentary rock record of the midwestern states, often thousands of feet in thickness, accumulated in these shifting marine and terrestrial environments. Lime-rich shells and skeletons of organisms as well as chemical precipitates from seawater also accumulated on the ancient sea floors. The layers of sediment were gradually compressed and hardened into limestone, dolomite, sandstone, and shale—Iowa's most common varieties of sedimentary bedrock.

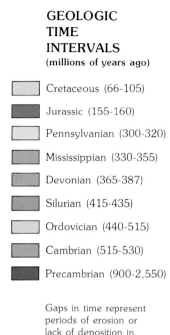

GEOLOGIC TIME INTERVALS
(millions of years ago)

Cretaceous (66-105)

Jurassic (155-160)

Pennsylvanian (300-320)

Mississippian (330-355)

Devonian (365-387)

Silurian (415-435)

Ordovician (440-515)

Cambrian (515-530)

Precambrian (900-2,550)

Gaps in time represent periods of erosion or lack of deposition in Iowa's rock record.

Remains of the diverse organisms that inhabited these environments were preserved as fossils in the hardening strata. Commonly found marine fossils include brachiopods, snails, crinoids, corals, trilobites, cephalopods, and fish. During arid periods when seaways were restricted and very saline, evaporation caused the accumulation of gypsum salts. At other times the incomplete decay of lush

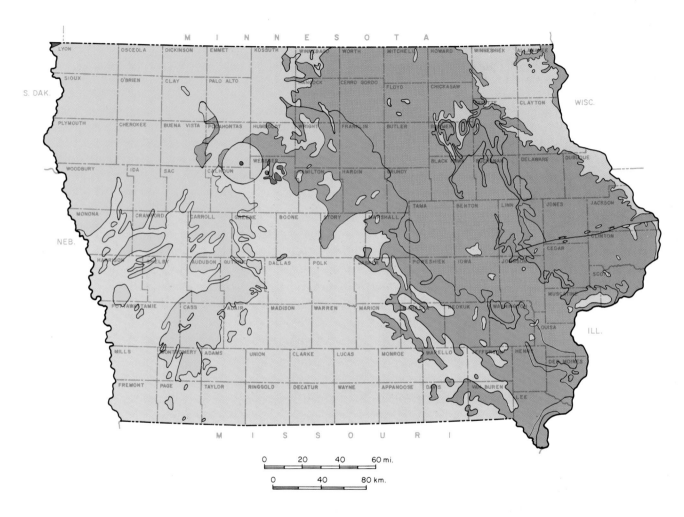

Bedrock Geology of Iowa

tropical vegetation growing along humid coastal swamps and river systems resulted in the preservation of carbon-rich deposits of coal and plant fossils. Limestone, dolomite, shale, gypsum, and coal are sedimentary rock resources that are mined and quarried in Iowa today.

With the passage of geologic time, the sedimentary rocks were warped, tilted, faulted, fractured, and partially eroded. Though Iowa today is situated on a very stable portion of the continent's interior, tremors originating elsewhere are occasionally felt here. The Plum River Fault Zone, though not a historically active fault, is evidence that more severe jolts once originated here, displacing and fracturing rocks across a portion of east-central Iowa.

Overall, this package of strata inclines downward to the southwest toward former sea basins in Kansas and Oklahoma. The result is that Iowa's bedrock, as it outcrops at various places across the state, represents not just the remains of the last great inland sea but a progression of marine occupations that began about 550 million years ago (Cambrian) and continued at least through 70 million years ago (Cretaceous)—an impressive interval of Earth history. A traveler starting out in western Iowa and continuing to the state's eastern and northern borders sees bedrock of increasingly older age. It is the southwesterly dip of the rock sequence, whose inclined layers were planed during later erosion, that keeps older and older strata within reach of the land surface. The banded appearance seen on the state's bedrock geology map on the previous page reflects this pattern as each deeper and older stratigraphic layer extends farther north-

east beyond the one above it.

The terrain that evolved on this sedimentary bedrock foundation is far from level or even. Deep, sheer-walled valleys separated by rolling upland plains, perhaps much like Iowa's northeastern counties look today, have been mapped statewide as drilling data from wells penetrating the glacial cover are compiled. Studies of the buried bedrock surface reveal a complex network of valleys. Seldom, however, do hills and valleys of the modern Iowa landscape coincide with these features of the buried ancestral landscape.

During much of the time sedimentary materials were being deposited, the North American continent was migrating at fingernail-growth speed northward toward its present latitudes. Today Iowa is landlocked, far removed from the oceans along the continent's borders. In considering how much geologic time has passed and how little of this buried rock record is visible in today's landscape, we come face to face with the widespread effects of another significant geologic event—glaciation, the "Ice Age" or Pleistocene Epoch.

Most of Iowa's landforms and the materials from which they have been molded are, geologically speaking, quite young. The Pleistocene spans the portion of geologic time from about 1.6 million years ago to the present (see stratigraphic key, p. 35). Most geologists regard the "Ice Age" as still in progress. The present climatic conditions are well within the range of other interglacial warmings that separated periods of ice advance. The term Quaternary is often used to refer to the Pleistocene, including its current interval of postglacial environmental change, the Holocene. These changing environments accompanied the arrival of human beings onto the Iowa landscape. Bands of Paleo-Indian hunters crossed the Bering Strait land bridge from Asia into North America about 12,000 years ago, and datable archaeological records of these people appear in Iowa's glacial-age deposits about 11,500 years ago—the first overlapping of the state's natural and cultural history.

When the drifting continental masses reached higher latitudes, glaciation became possible in the northern hemisphere. Evidence of earlier, southern hemisphere glaciations also exists in the geologic record, especially from Pennsylvanian-age rocks, which record numerous sea-level fluctuations about 300 million years ago. The Pleistocene glacial expansions were triggered by shifts in the Earth's climatic heat balance. Though no single cause is apparent, the changes in seasonal air-mass positions necessary to increase snowfall and reduce melting have been attributed to variations in solar energy caused by the wobbling motion of the Earth's axis and variations in the geometry of the Earth's orbit around the Sun. Colder and wetter climates produced persistent snowfall and dense ice accumulation in the Canadian Arctic.

The great weight of the thickening ice caused it to become mobile and flow southward, eventually into more moderate climates. Ice sheets thousands of feet thick spread across North America from the Atlantic seaboard to the plains east of the Rocky Mountains. At maximum extent, the glacial ice may have been as much as two miles thick as far south as the Great Lakes' northern shores. Along the way these continental glaciers subdivided into ice lobes as they encountered regional variations in the underlying topography or as the mechanisms of ice flow changed. The tremendous weight of the ice sheets actually depressed the Earth's crust, and later, as the ice melted, the crust began rebounding to its former position. The results of this movement can be seen and measured today in the vicinity of the Great Lakes and Hudson Bay where former shorelines are now elevated. The growth of these glaciers also took enormous amounts of water out of circulation and lowered the sea level as much as three hundred feet along the world's coastlines.

Much of the knowledge of the tempo of Pleistocene events worldwide is derived from evidence preserved in ocean basins. These depths, the planet's geological pockets, contain the most continuous record of Pleistocene sedimentation available for study. Cycles of glaciation show up in cores of deep-sea muds, deposits which register changes in seawater chemistry (oxygen isotopes) and marine life, especially tiny protozoans (foraminifera) that are particularly sensitive to temperature. On the continent, on the other hand, we are left to puzzle over the details of a complex but incomplete record of the passage of Pleistocene time. The spectacular glacial events themselves occupied relatively brief periods, though they made a significant impact in a short time. Weathering, soil development, and erosion were the more common long-term geological processes during the Pleistocene. Geologists are continually working to interpret and integrate the information from glacial-age deposits left in geographic settings throughout the world. Since the most

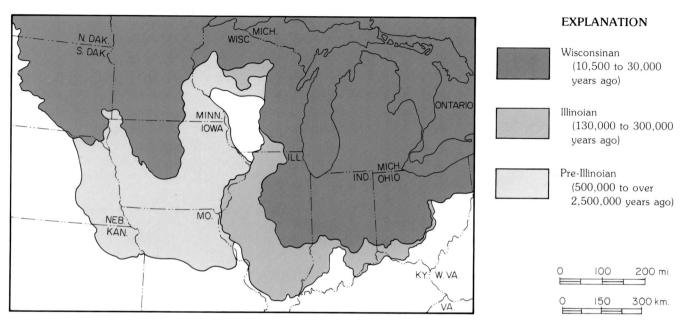

Limits of Major Glacial Advances in the Upper Midwest

EXPLANATION

Wisconsinan
(10,500 to 30,000
years ago)

Illinoian
(130,000 to 300,000
years ago)

Pre-Illinoian
(500,000 to over
2,500,000 years ago)

recent events leave the most accessible record, our clearest focus on the Midwest's and Iowa's glacial past applies to a short and relatively recent interval of geologic time.

During the Pleistocene, continental glaciers repeatedly advanced over all or parts of Iowa, each advance subtracting some of the earlier deposits and adding its own depositional record before melting away. The principal periods of ice cover, from oldest to youngest, are grouped into the Pre-Illinoian, Illinoian, and Wisconsinan glacial stages. The intervals of climatic warming that separated these glacial periods are referred to as interglacial stages, and the two youngest and best known are the Yarmouth and Sangamon. The geographic limits of these various glacial expansions into the central United States are shown on the map to the left. Because the margins of continental glaciers were irregularly active, the areas shown as covered during these ice advances are composite views, combining areas occupied throughout a range of time.

The idea that huge glaciers once covered North America and northern Europe did not become a serious scientific theory until about 1840 to 1860, when evidence was published by the renowned naturalist Louis Agassiz. About a century ago the state of Iowa played an important role in presenting the stratigraphic facts which established the concept of multiple glacial periods in North America during the Pleistocene. The separation of the "Ice Age" into a series of glacial episodes was possible here because of Iowa's geographic position with respect to coverage by the various ice sheets and because of the amount of the Pleistocene record preserved in the state.

Unraveling Iowa's geologic evidence for multiple glacial advances has established concepts and terminology used throughout North America. The Pre-Illinoian stages of the early Pleistocene, once referred to as the "Nebraskan" and "Kansan" separated by the "Aftonian" interglacial stage, were defined from evidence in eastern Nebraska and southwestern Iowa, particularly the Afton-Thayer area of Union County. The complexity of these glacial periods, including the existence of warm interglacial episodes as interpreted from the "Aftonian" gravels and their classic fauna of Pleistocene mammals as well as buried soils or "gumbotils," was presented by Thomas C. Chamberlin, William J. McGee, Samuel Calvin, Bohumil Shimek, H. Foster Bain, Frank Leverett, and George F. Kay. Other important Iowa contributions to North American Pleistocene nomenclature include the type-locality of the Yarmouth interglacial stage, named for the town of Yarmouth in Des Moines County, and the type-section of the widespread Loveland Loess, named for the town of Loveland in Pottawattamie County. In addition, the windblown origin of loess, based on study of land snails, was first confirmed by Shimek in 1895. Research into Iowa's glacial deposits remains important to the evolution of geologic thought and to studies that span the range of Quaternary events, environments, and biota on the North American continent.

The various glacial advances that spread slowly across Iowa's landscape carried within them frozen soil and rocks gouged and scooped from along their routes across more northern landscapes. Some of the glacial load lodged against the ground, pressed beneath the weight of the advancing ice. This grayish clay with an assortment of sand, pebbles, and boulders scattered throughout was deposited directly by glacial ice and is termed till. Most of the remaining glacial load was released as melt-out debris when glaciers began to waste away. In these environments of lingering, disintegrating glacial ice, deposits were frequently rearranged by mudflows, slumping, deformation, collapse, and meltwater. Diamicton is an increasingly popular term used to describe poorly sorted glacial deposits including till and these reworked deposits. Stagnating glaciers also left behind ice-contact deposits that were actually shaped by the ice's contours and edges. Meltwater flowing away

Timothy J. Kemmis

Jean C. Prior

Brian J. Witzke

Glacial drift *(till shown): deposits of clay, silt, sand, and cobbles left by glaciers or their meltwater streams.*

Eolian sand: *upland deposit of windblown sand.*

Loess: *deposit of windblown silt, with minor amounts of sand and clay.*

22

Timothy J. Kemmis

Timothy J. Kemmis

Brian J. Witzke

Alluvium: *clay, silt, sand, or gravel deposited by rivers.*

Paleosol: *a fossil soil, usually buried; indicates weathering of an earlier, stable land surface.*

Bedrock *(limestone shown): rock strata exposed at the land surface or occurring beneath younger soils, glacial deposits, or alluvium.*

23

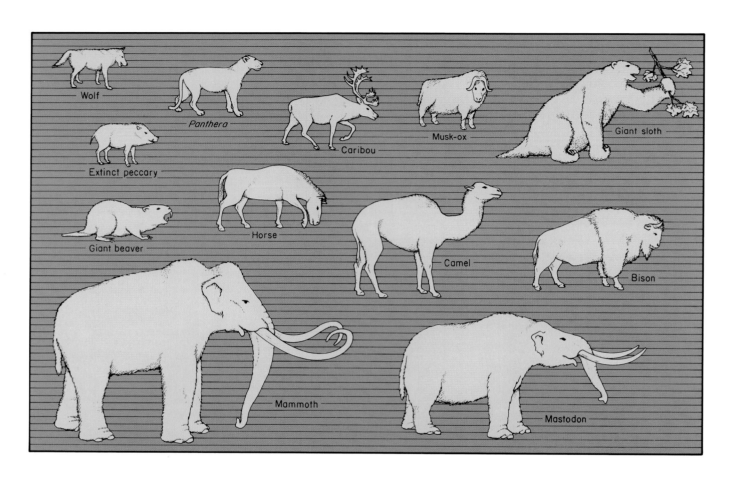

Fossils of large vertebrates that lived in Iowa during the Pleistocene period are found in the state's glacial and interglacial deposits. All animals are shown in approximate scale to each other.

from the ice front left outwash deposits of sorted sand and gravel. Layers of silt, clay, and peat accumulated in quieter lakes and ponds on the ice and along its margins. An old but useful term that encompasses all of these types of glacial and meltwater deposits is drift. These deposits unevenly blanket almost all of Iowa's bedrock foundation; while there is no drift in parts of northeastern Iowa, it ranges in thickness to over 600 feet in the west-central part of the state.

As a side effect of glaciation, persistent winds, driven by marked temperature contrasts between ice-covered and ice-free areas, carried aloft pulverized glacial debris concentrated in river valleys and left behind eolian deposits of loose, gritty yellowish silt (loess) and fine sand. These windblown deposits, while not derived directly from glacial action, are commonly present in glaciated landscapes. Within the deposits of loess and drift are color and grain-size differences caused by later soil-forming processes across the landscape. These paleosols or ancient soils are important markers indicating periods of landscape stability and development of vegetation during the Pleistocene. All of these materials, derived in some way from the presence of glaciers in or near Iowa, are basic components of today's landscape (photos, pp. 22–23).

Farmers in many areas of Iowa have cleared boulders from their fields only to find a season or two later that others have worked their way to the surface during winter's freeze and thaw. These igneous and metamorphic rocks are conspicuous geologic strangers here where sedimentary rocks dominate—hence the name glacial erratics. These travel-worn fieldstones can be traced to outcrops of igneous and metamorphic rocks in Minnesota, Wisconsin, Michigan, and Canada. Their presence in Iowa offers further evidence of the power and direction of flow of Pleistocene ice sheets.

The advancing glaciers left other, more detailed tracks of their activity. Glacial grooves are sometimes observed where the bedrock surface has been uncovered by stream erosion, quarrying, or construction activities. These parallel grooves were etched into underlying strata as glacial ice inched across the rock surface, dragging cobbles and boulders at its base. The direction of local ice movement can be determined from compass bearings taken along the grooves. The only glacial grooves permanently on view in Iowa are at the Stainbrook Geological Preserve in northern Johnson County. Finely etched grooves and scratches or striations as well as faceted surfaces are commonly seen on individual pebbles and cobbles carried by the ice.

Many ancient Iowa stream valleys were obliterated by overriding ice and their waters rearranged and diverted to form new valleys. Glacial drift filled the old valleys and buried former riverbeds far beneath the present land surface. New streams were born from melting ice margins, and they transported enormous quantities of water and sediment. River valleys that received glacial meltwaters and produced loess also collected and stored large amounts of coarse glacial debris—sand and gravel, another important natural resource in Iowa. As these river volumes and velocities changed with the seasons, additional layers of clay and silt were deposited along the

Molars of a young mastodon (left) and mammoth (middle), 10,000 to 15,000 years old, are compared with the tooth from a young horse (right). Teeth shown at 39 percent of original size.

valley floodplains. All these water-sorted materials are called alluvium or alluvial deposits. Some of the state's valleys today seem unusually large for the size of the rivers now occupying them. These underfit streams are good indicators of the great volumes of meltwater and sediment that enlarged and partially backfilled these valleys during the Pleistocene.

As the continental ice sheets waxed and waned over Iowa and other midwestern landscapes, distinct shifts in the distribution of plant and animal life followed. The fossil remains of these various organisms permit glimpses into Iowa's climate and environment during different phases of Quaternary history. Radiocarbon (carbon-14) dating, useful for materials less than 45,000 years old, can be applied to organic remains such as fossil wood, peat, plants, bones, shells, and charcoal.

Large mammals that inhabited Iowa during this fascinating phase of landscape history include horse, camel,

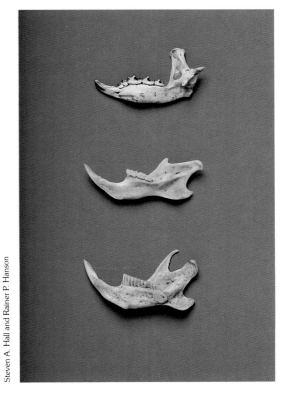

Differences in the tiny teeth of these Delaware County rodent fossils, the flesh-eating short-tailed shrew (top), browsing deer mouse (middle), and grass-eating woodland vole (bottom), are indicators of environmental conditions 6,000 years ago. Jaws shown at 220 percent of original size.

llama, giant beaver, caribou, reindeer, musk-ox, giant ground sloth, elk, and bison (drawing, p. 24). With their enormous teeth and tusks, mastodons and woolly mammoths, which now are extinct but resembled modern elephants, are two of the better-known large mammals that roamed North America during glacial and interglacial stages. In Iowa, specimens of bones and teeth from these great beasts have been documented from all ninety-nine counties, many from sand-and-gravel deposits along the state's river valleys (photo, far left). Differences in tooth structure enable vertebrate paleontologists to identify these creatures and interpret their surrounding environments. For example, a mammoth's molar is composed of a series of laminated plates, having a gently ridged surface. This large grinding surface indicates mammoths were grazing animals with a probable diet of grasses. The tooth of a mastodon, on the other hand, has numerous peaks or cusps, well adapted to browsing in timber for branches, twigs, cones, and leaves.

The fossils of small mammals are particularly valuable as indicators of environmental and climatic change through time. Teeth and jaws of various species of fox, hare, vole, shrew, and lemming reveal whether they were adapted to grasslands, forest, or tundra (photo, left).

Vegetation characteristic of the open tundra and dense boreal spruce forests that grow today in the colder latitudes of the northern United States, Canada, and Arctic regions extended with the glacial advances to more southerly ranges. Studies show that the period of most intense glacial cold during the late-Wisconsinan occurred in Iowa between 16,500 and 21,000 years ago, when the

Fossil seeds are used to interpret past local vegetation. The black seed of dockleaf smartweed grew about 34,000 years ago along a marsh edge in Warren County. For comparison, a modern specimen is shown with its outer yellow covering.

climate was wetter and July temperatures were about 20 to 23 degrees colder than today. This period included a rich biota of arctic and subarctic species that indicated tundra conditions and closely resembled the environment at the modern tree limit in northern Canada. Postglacial warming brought a mixture of coniferous and deciduous forest species interspersed with grasslands. This complex environment with its diverse flora and fauna has no current counterpart among the world's vegetation communities and indicates that quite different ecological

niches were available during the late Pleistocene than exist now. Oak savanna and prairie eventually became the major presettlement vegetation communities in Iowa. These shifts in climates and habitats and accompanying flora and fauna are documented by paleobotanists, paleontologists, and entomologists who study fossil plants (photo, left), pollen grains, snails, clams, vertebrates, and insect remains embedded in the peats and organic-rich sediments that accumulated in Pleistocene wetlands.

In our present interglacial environment, the principal geologic agent shaping the Iowa landscape is flowing water. Uplands are being lowered in elevation, slopes are steepening in some places and leveling out in others, and lowlands are accumulating sediment brought by surface runoff and stream flow. Rainfall and snowmelt trickle downslope and either soak into the ground, evaporate, or eventually are collected by one of Iowa's many streams. Water from streams and lakes evaporates into the atmosphere where it is available again as potential precipitation, thus completing the hydrologic cycle. Groundwater, another component of this cycle, moves through soil and rock along shallow local flow paths or into deeper regional systems. These sources of groundwater, or aquifers, are present within river alluvium, glacial drift, porous sandstone, and creviced limestone deposits beneath Iowa's landscape.

Emerging from the great lengths of geologic time and the changes brought by diverse geological events are the landscapes of Iowa that are known to us today. Only one other factor has altered the land's appearance—the activities of people. Before Euro-American settlement

Iowa's present climate supported a vast native prairie prior to the time of Euro-American settlement. A remnant of this original vegetation is protected at Williams Prairie State Preserve in Johnson County. This August view of the moist meadow in bloom shows lavender spikes of blazing star, white tufts of rattlesnake master, and golden rays of the black-eyed susan.

of Iowa, much of the land was covered with prairie vegetation. Parts of the state were dotted with bogs and marshes, as well as other wetlands known as fens. The original vegetation cloaked the state in what was described by the earliest pioneers as a sea of grass. Fingers of forest were confined along river valleys and in the extreme northeastern part of the state. Today only a few remnants, less than 0.1 percent, of the original prairie remain (photo, above). Many of the native wetlands, including fens, bogs, marshes, and lakes, have been drained. Much of the land surface is cultivated. The root systems and organisms of Iowa's past vegetation communities had transformed the raw deposits of loess, glacial drift, and valley alluvium into the state's invaluable inches of rich topsoil. Timely rains, accessibility of the low-relief, glaciated landscapes to farm machinery, and the natural ability of the land to grow grasses make Iowa a leading agricultural state, particularly in the production of grain crops such as corn and soybeans.

Iowa's landscapes and landforms are the direct result of a wide variety of Quaternary processes. These processes, operating at different times and intensities over different parts of the state, and the materials and shapes they left behind bear primary responsibility for the appearance of Iowa's landscape today.

Landform Regions

Iowa contains 55,986 square miles. The eastern border with Illinois and Wisconsin is formed by the Mississippi River. The western border with Nebraska and South Dakota follows the course of the Missouri and Big Sioux rivers respectively. The northern border with Minnesota nearly coincides with parallel 43°30′ north latitude, while the southern border with Missouri is an arbitrated line that approximates the arc of parallel 40°35′ north latitude eastward to the Des Moines River and then follows that river's course southeast to the Mississippi.

The rolling, predominantly agricultural landscape is generally characterized by low relief, fertile soils, numerous rivers, and a variety of glacial deposits mantling gently inclined layers of sedimentary bedrock. The state's highest elevation, 1,670 feet above sea level, is located northeast of Sibley in Osceola County (NE 1/4, Sec. 29, T100N, R41W). The state's lowest elevation, 480 feet above sea level, is located southwest of Keokuk at the confluence of the Des Moines and Mississippi rivers in southeastern Iowa (SE 1/4, Sec. 34, T65N, R5W).

Iowa's landscapes are remarkably diverse. The map on page 31 outlines the state's seven topographic regions: the Des Moines Lobe, the Loess Hills, the Southern Iowa Drift Plain, the Iowan Surface, the Northwest Iowa Plains, the Paleozoic Plateau, and the Alluvial Plains. These regions are distinguished on the basis of physical appearance, and their observable differences result from variations in geologic history—an intriguing combination of time, earth materials, and events. As different portions of Iowa were released from the grasp of various glacial advances, stream erosion took over and continued to shape the land. Each region contains distinct landscape patterns and features that resulted from erosional activity at different times, in varying intensity, into variable deposits of loess, drift, alluvium, or bedrock. Some regions contrast sharply, with an obvious topographic boundary separating them. Other boundaries are less clear, and the change from one landscape pattern to another may occur gradually over several miles.

A series of maps follow containing statewide information on rivers and lakes, topographic relief, and terrain characteristics and age. Each landform region is then discussed individually in terms of its typical landscape appearance, underlying earth materials, and geologic origins. The sequence of chapters is intended to introduce information about geologic events and deposits in a progression useful to understanding a specific region while also building an appreciation for the combined history that links the regions together.

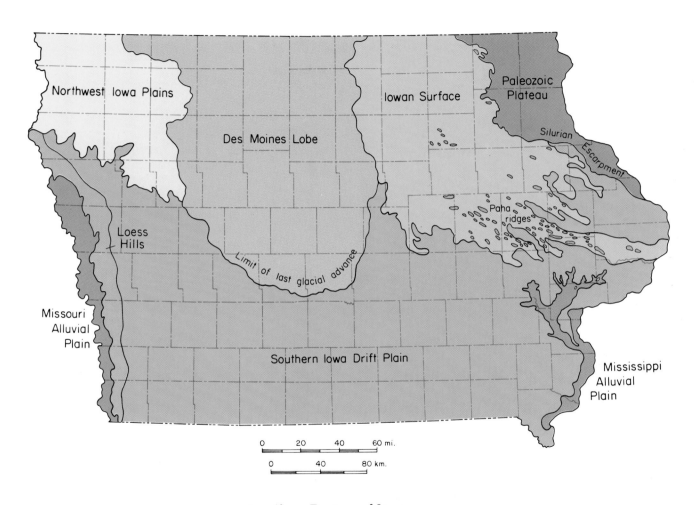

Northwest Iowa Plains

Iowan Surface

Paleozoic Plateau

Des Moines Lobe

Silurian

Escarpment

Paha ridges

Loess Hills

Limit of last glacial advance

Missouri Alluvial Plain

Southern Iowa Drift Plain

Mississippi Alluvial Plain

0   20   40   60 mi.

0   40   80 km.

Landform Regions of Iowa

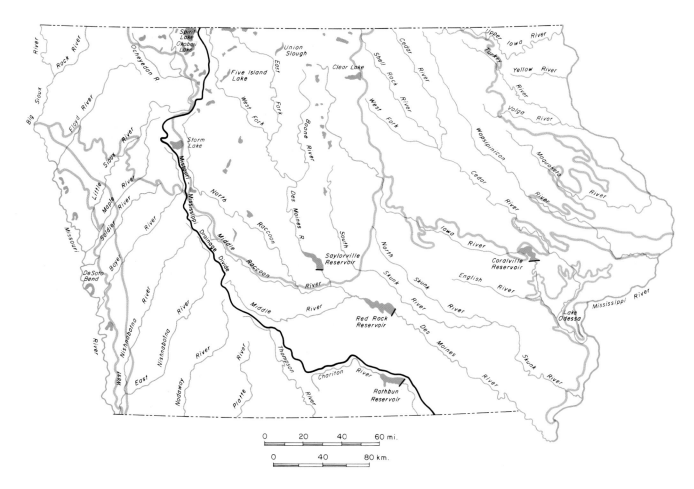

Rivers and Lakes of Iowa

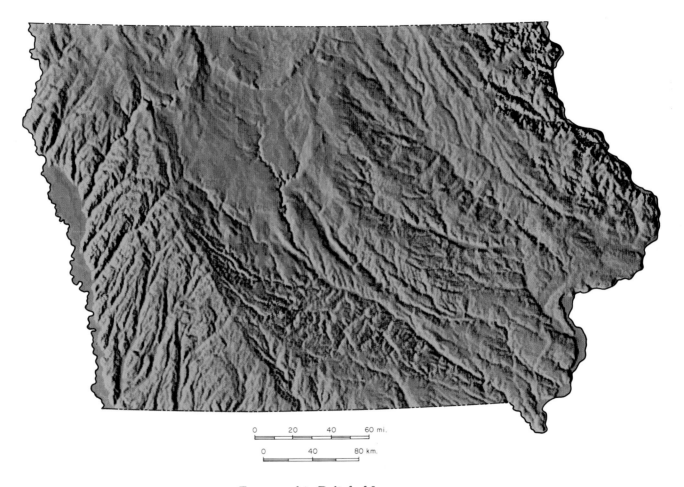

0   20   40   60 mi.

0   40   80 km.

Topographic Relief of Iowa

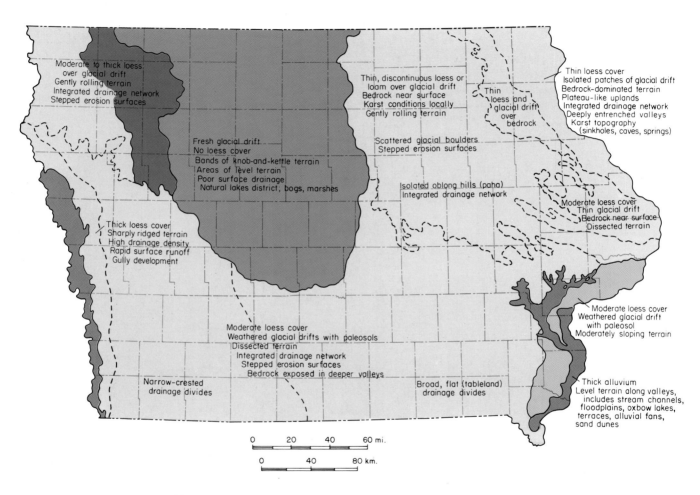

Moderate to thick loess
over glacial drift
Gently rolling terrain
Integrated drainage network
Stepped erosion surfaces

Thin, discontinuous loess or
loam over glacial drift
Bedrock near surface
Karst conditions locally
Gently rolling terrain

Thin
loess and
glacial drift
over
bedrock

Thin loess cover
Isolated patches of glacial drift
Bedrock-dominated terrain
Plateau-like uplands
Integrated drainage network
Deeply entrenched valleys
Karst topography
(sinkholes, caves, springs)

Fresh glacial drift
No loess cover
Bands of knob-and-kettle terrain
Areas of level terrain
Poor surface drainage
Natural lakes district; bogs, marshes

Scattered glacial boulders
Stepped erosion surfaces

Isolated oblong hills (paha)
Integrated drainage network

Moderate loess cover
Thin glacial drift
Bedrock near surface
Dissected terrain

Thick loess cover
Sharply ridged terrain
High drainage density
Rapid surface runoff
Gully development

Moderate loess cover
Weathered glacial drift
with paleosol
Moderately sloping terrain

Moderate loess cover
Weathered glacial drifts with paleosols
Dissected terrain
Integrated drainage network
Stepped erosion surfaces
Bedrock exposed in deeper valleys

Narrow-crested
drainage divides

Broad, flat (tableland)
drainage divides

Thick alluvium
Level terrain along valleys,
includes stream channels,
floodplains, oxbow lakes,
terraces, alluvial fans,
sand dunes

0   20   40   60 mi.

0   40   80 km.

Landform Materials and Terrain Characteristics of Iowa

34

		Holocene Stage 　　DeForest Formation—alluvium* 　　Eolian sand (locally)	Present
			10,500 years ago
		Wisconsinan Glacial Stage 　　Dows Formation—glacial drift 　　Wisconsinan loesses and eolian sand	12,500-14,000 12,500-31,000
		Sheldon Creek Formation—glacial drift	20,000-30,000 ?
		Sangamonian Interglacial Stage 　　Sangamon Soil (paleosol) and alluvium	
			130,000
		Illinoian Glacial Stage 　　Glasford Formation (Kellerville Till Member)—glacial drift	
			300,000
		Yarmouthian Interglacial Stage 　　Yarmouth Soil (paleosol) and alluvium	
			500,000
		Pre-Illinoian Glacial and Interglacial Stages (numerous) 　　Wolf Creek Formation—glacial drifts	
		- -	700,000
		Alburnett Formation—glacial drifts	1,650,000
		Older glacial and interglacial deposits	
			2,500,000

QUATERNARY PERIOD — Pleistocene Epoch

TERTIARY ? — Pliocene ?

Accurate proportions for time intervals on the left

*See photos of basic earth materials, pages 22-23.

Stratigraphic Key to Landform Materials

Des Moines Lobe

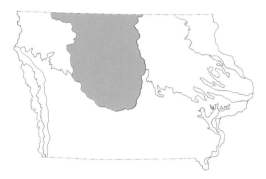

The hills . . . pitch toward every
point of the compass, they are of
every height and shape, . . . they
enclose anon high tablelands, anon
wide low valleys that open nowhere;
they carry lakes on their summits and
undrained marshes at their feet; their
gentler slopes are beautiful prairies
easily amenable to the plough, their
crowns are often beds of gravel
capped with bowlders and reefs
of driven sand.

—Thomas H. Macbride
"Geology of Osceola and
Dickinson Counties," 1900

The icy grip of continental glaciers was one of the most significant geologic processes to affect the Iowa landscape. Most of the deposits underlying today's land surface are composed of materials known as drift that were moved here by glaciers. The arrival of these glaciers in the state began over two million years ago, and numerous reappearances are recorded in the deposits they left behind. For all of this massive effort, however, only the landscapes of north-central Iowa still display the actual shapes that resulted directly from glacial action. This region, known as the Des Moines Lobe, is the part of the state last touched by the huge sheets of frozen water that invaded Iowa in the past. This last glacial episode occurred only 12,000 to 14,000 years ago.

The relatively recent date of this geologic event, as compared to older glacial episodes, provides a more detailed record of its existence as well as a much clearer picture of how the glacier behaved while it was here. In recent years geologists have mapped and studied the details of the Des Moines Lobe landforms, including not only their geographic distribution but the grain size, mineral content, and internal structures of their underlying deposits. This important work advances our understanding of glacial processes and their contribution to the evolution of north-central Iowa landscapes. In addition, such research provides valuable technical information used to predict the engineering, agricultural, and hydrologic properties of these deposits. This information has direct application for those who construct buildings, manage farms, need to locate groundwater supplies, or evaluate waste-disposal sites. These detailed studies reveal a complex

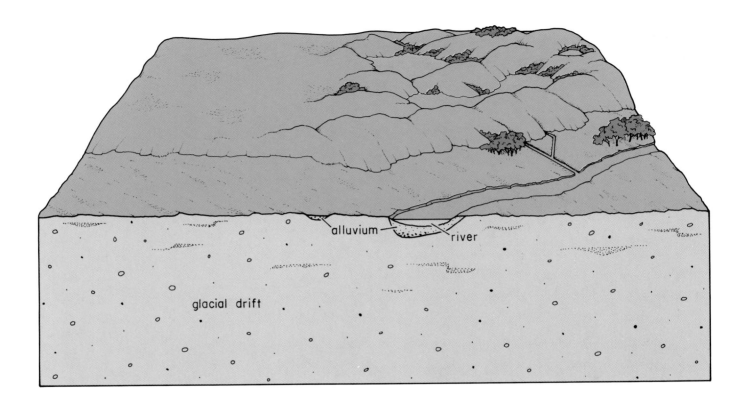

alluvium

river

glacial drift

but systematic assemblage of landforms that reflect dynamic environments of glacial activity through time across different parts of the Des Moines Lobe.

The picture of glacial activity that emerges is one of a surging ice front rapidly expanding into Iowa from a center of activity along the southern margin of the vast Wisconsinan ice sheet that lay across North America (see map, p. 20). The Des Moines Lobe advanced through North and South Dakota and Minnesota and into north-central Iowa, halting at what is now the city of Des Moines and in the process establishing the present course of the Raccoon River. Today, the Iowa portion of the Des

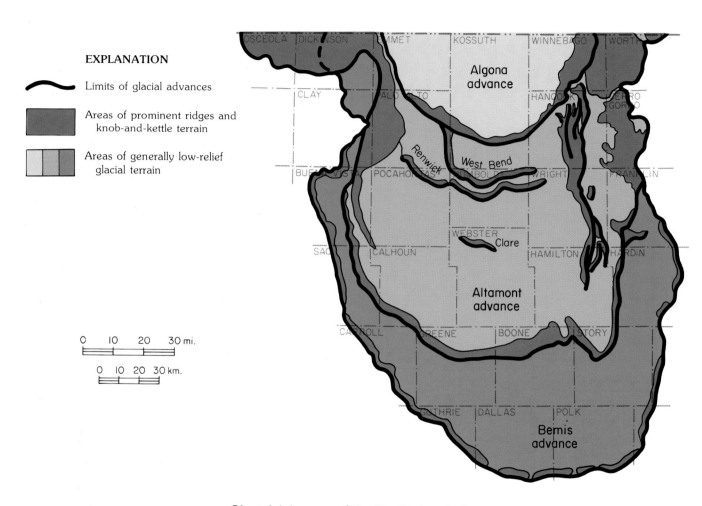

EXPLANATION

~~~~ Limits of glacial advances

■ Areas of prominent ridges and knob-and-kettle terrain

▨ Areas of generally low-relief glacial terrain

0   10   20   30 mi.

0   10   20   30 km.

Algona advance

Renwick

West Bend

Clare

Altamont advance

Bemis advance

Glacial Advances of the Des Moines Lobe

Moines Lobe is further outlined by the cities of Clear Lake, Eldora, Carroll, Storm Lake, and West Okoboji. The glacier's advance occurred well after the period of maximum glacial cold (16,500 to 21,000 years ago) at these latitudes, and the moderating climatic conditions were hospitable to a richly diverse flora and fauna that inhabited meadows and woodlands along the ice margins. This episode of glacial activity, composed of at least three major surges, was followed by large-scale stagnation of the glacial ice. The process of decay and disintegration was the dominant method of ice "retreat" over large areas of the Des Moines Lobe.

A detailed chronology of the occupation of north-central Iowa by Des Moines Lobe ice is possible because of radiocarbon dating of peat from bogs as well as logs and stumps of the spruce, hemlock, and larch that were uprooted by the advancing ice. Abundant organic-rich sediment and wood were collected at the contacts between different underlying deposits and from within the Lobe's glacial drift. While the shrunken, relatively inactive remnants of North America's last continental ice sheet reside today in Greenland, the behavior of more active valley glaciers can be examined in places such as Alaska and Scandinavia. Here geologists can study modern glacial processes in an attempt to better understand the origins of glacial deposits thousands of years old elsewhere in the world. Though these valley glaciers are confined by steep mountainsides, valuable comparisons still can be made between their flow behavior and environments of sedimentation and those of glaciers in Iowa's geologic past.

The most prominent landform patterns observed on the Des Moines Lobe are related to the major end moraines (map, left). These curved, concentric bands of ridges and hummocky terrain outline the maximum extent of active ice advances and the position of prolonged, stationary margins of stagnant ice masses. The deposits marooned around these ice margins became well-defined features of the land surface once the ice melted away. Their pattern of alignment, parallel with the outline of the Lobe, shows their origins were closely associated with the changing ice margins. The limits of three major glacial advances are marked by the Bemis, Altamont, and Algona end moraines. Within the Altamont surge, minor advances are also recorded by the Clare, Renwick, and West Bend end moraines. Their names are taken from localities where their deposits are best displayed: Bemis and Altamont are towns in South Dakota while Clare, Renwick, West Bend, and Algona are north-central Iowa communities.

Landforms and deposits vary along these former ice margins. The portions of morainal ridges extending east-west across the interior portion of the Lobe appear as smooth, prominent, south-facing escarpments, as much as 120 feet high. They formed along the central axis of the surging glacier, where the ice was especially mobile. This region of active oscillations of the ice front had a relatively uncomplicated history of melting. The process left behind thick deposits of compact, uniform pebbly loam called glacial till and generally produced low-relief landscapes punctuated by distinct ridges marking the limits of major ice advances. These ridge fronts are easily seen where the Altamont end moraine crosses Greene and western Boone counties and farther north where the younger Algona end moraine crosses eastern Palo Alto, Kossuth, and western Hancock counties.

As these morainal ridges begin to curve along the lateral margins of the Lobe into a more north-south orientation toward the Iowa-Minnesota border, they grade, sometimes abruptly, into broader belts of rough hummocky topography, especially in the northwestern and northeastern portions of the Lobe. This rumpled terrain, also called knob-and-kettle topography, is particularly prominent in Dickinson and adjacent parts of Emmet, Palo Alto, and Clay counties and, on the opposite side of the Lobe, in eastern Winnebago and central Hancock and Cerro Gordo counties.

These landscapes suggest that the lateral margins of the ice lobe moved more slowly and caused the ice flow to become tightly compressed. As a result, the debris normally transported near the glacier's base was forced upward along zones of shearing fracture and flow within the ice and eventually melted out onto the ice surface. There the glacier's load sometimes mixed with meltwater to form mud flows; it was also redeposited in meltwater channels and pools on the ice surface; and it was subject to slumping, deformation, and collapse as the underlying ice slowly disintegrated. The resulting deposits are composed of the usual dense pebbly loam near the bottom. This foundation of basal till, however, is veneered with stony heaps of uneven hummocky deposits that may include sandy loams, water-sorted sands and gravels, or lake-deposited silts and clays—all of which can change

quickly over short vertical and horizontal distances. The jumbled hills, ridges, and swales further suggest that this slow-moving or stagnant glacial ice was continuously present in this portion of the Des Moines Lobe. Also, some of these landform patterns are composites; they combine the topographic effects of the earlier Bemis ice advance with those of the overriding Altamont and Algona ice advances.

As the stagnant ice wasted away from north-central Iowa following the various surge advances, cavities and tunnels formed within and beneath the ice, some of which were linked to the glacier's surface. These ice-walled conduits served as drainageways for meltwater and glacial debris. The larger deeper portions of these networks were probably open throughout the year, while the smaller upper portions were frozen shut and redeveloped each spring in new locations. In time these drainage systems collapsed and disintegrated, resulting in hummocky topography. The larger, more permanent segments of the systems became sites of the Lobe's present-day rivers and some of its larger lakes. The collapse of

Prairie potholes dot the landscape of Doolittle Prairie State Preserve in Story County. These saucer-shaped wetlands formed during stagnation and melting of the Altamont ice advance about 13,000 years ago. The subtle drainage links between them, indicated by soil moisture, vegetation, and land use, show they formed along meltwater routes.

Photo by Gary Hightshoe

smaller segments of the karst-like systems in the ice led to the development of shallower upland swales and depressions. The irregular concentration of glacial debris that did not get flushed from these networks resulted in saddle-like ridges which separated segments of the system into kettle-like basins. Many of the smaller kettle lakes, prairie potholes, and other depressions and lowlands on the Des Moines Lobe actually display a subtly linked surface drainage pattern (photo, p. 41). One of the most significant and unexpected findings to emerge from recent detailed mapping of the maze of landforms and deposits across the Des Moines Lobe is this pattern of connected drainage routes.

Other local characteristics of the lingering ice mass contributed to the irregular pattern of landforms and deposits. Many landform features inherited from patterns and processes of ice decay remain visible on the land surface. For example, some large crevices and chambers within the ice that were open to the glacier's melting surface became filled with water-transported deposits of sand and gravel. These gravelly fillings exist on today's landscape as isolated, roughly conical hills called kames. Ocheyedan Mound State Preserve in Osceola County (photo, above), Pilot Knob State Park in Hancock County, and Pilot Mound State Forest in Boone County are excellent examples of kames. Similarly, the courses of streams confined in tunnels at the base of the ice sometimes clogged with sand and gravel. These alluvial-filled channels, relics of rivers that flowed beneath the ice, are seen today as narrow, winding ridges called eskers. While rare in Iowa, both Cayler Prairie State Preserve and Koppen Prairie near Grovers Lake in Dickinson County exhibit the crooked ridge forms marking the traces of these subglacial drainageways.

The reverse of these patterns is seen in the steep-sided, bowl-shaped depressions called kettles which dimple the present land surface. Kettles mark the positions of large, isolated blocks of relatively clean glacial ice which melted slowly and perhaps were buried and protected by a thin cover of soil and rock rubble. The impressive, crater-like form of the Freda Haffner Kettlehole, a state preserve in Dickinson County, developed by this process (photo, p. 45). These features have no surface drainage outlets. Organic deposits that have accumulated in the bottoms of closed depressions include microscopic grains of pollen from the vegetation that has covered the kettle's sideslopes and surroundings during the geologic past. Examination of this fossil material can be used to reconstruct the sequence of plant cover and related climatic change in the surrounding geographic area from the time the ice melted to the present.

It is important to realize that Iowa's end moraines are not uniform or consistent in their appearance or composition because so many different environments of deposition existed along a single ice border. Consequently, these features do not conform to the classic "recessional" moraines caused by active and repeated glacial advance and retreat, which was more common in states to the north and east of Iowa and previously was used to explain the origin of Des Moines Lobe moraines. Instead, we see rapid glacial advances into a moderating climate in north-central Iowa, followed by regional stagnation of the ice—a picture more consistent with surge-type glacial activity. The two broad categories of "ground moraine" and "end moraine" previously mapped in Iowa are not adequate to describe what really happened here, nor do they account for the variety of landforms and deposits that remain.

Between these various ice-marginal landscapes, other features and deposits have been identified. For example, behind the morainal ridges, numerous subtle sets of narrow, parallel ridge systems sometimes occur over large areas. These low-relief, slightly irregular landscapes are described as having "washboard" or "swell and swale" topography. Such minor-moraine crests, which are spaced about 350 feet apart and may rise no more than 5 feet, are barely perceptible on the ground; however, their patterns are clearly visible in aerial photographs, such as those of Story and Hardin counties.

Other identified features include sites of once large but now extinct glacial lakes, which are seen today as broad flats covering hundreds of square miles in southern Kossuth and Hancock counties (glacial Lake Jones) and also in a band through Wright, Hamilton, and southern Webster counties (glacial Lake Wright). In addition, extensive outwash deposits of sand and gravel carried and sorted by meltwater floods are found in front of some of the prominent end moraines along former overflow channels and glacial meltwater routes that are now the courses of such rivers as the Cedar, Shell Rock, Iowa, Skunk, Des Moines, Raccoon, Boyer, and Little Sioux. In fact, the Little Sioux River basin doubled in size following the Des Moines Lobe glacial blockage of Mississippi-bound rivers in northwest Iowa. This diverted drainage caused the temporary filling of another glacial lake (Lake Spencer) along the ice margin and the catastrophic overflow, downcutting, and permanent rerouting of these waters into the Missouri drainage system.

There are only a handful of major valleys on the Des Moines Lobe, and these are deep and narrow owing to their rapid excavation by swift meltwaters. The Des Moines River, in the most prominent of these valleys, flows roughly down the axis of the Des Moines Lobe landform region. The steep-sided valley, cut into Pennsylvanian bedrock, is dammed to hold back some of the river's flow, forming Saylorville Reservoir, the only major artificial lake on the Des Moines Lobe. Some uneroded remnants of the valley's outwash deposits occur as terraces perched along the valley sides. These features are often sites of commercial sand-and-gravel production, both on and off the Lobe. Terraces also contain important geologic records of successive episodes of postglacial erosion and deposition which occurred in response to melting of ice from north-central Iowa. The valley of Brushy Creek, a tributary of the Des Moines River in Webster County, hosts a prominent series of these topographic features spanning the entire evolution of the valley from over 11,000 years ago to the period of Euro-American settlement about 150 years ago. These young valleys are the only large-scale erosional features to be seen on the Des Moines Lobe. All other major landscape features were constructed by glacial ice or some related activity and thus are quite different from landscape shapes seen elsewhere in Iowa.

Nearly all of Iowa's natural lakes occur on the Des Moines Lobe, and they are noticeably clustered among the knobby hills close to the Minnesota border (see map, p. 32). Storm Lake, Lake Okoboji, Spirit Lake, Clear Lake, and numerous smaller ponds, sloughs, and bogs are characteristic of young, postglacial landscapes and their sluggish, inefficient drainage networks. The Iowa "Great Lakes" have been a popular vacationing area since the middle of the nineteenth century. These lakes, along with ponds and marshes sometimes referred to as prairie potholes, provide valuable wildlife habitat (photo, p. 46). The historic attraction of these wetlands for great flocks of nesting and migrating waterfowl is reflected in the names given to the Des Moines Lobe towns of Mallard, Plover, and Curlew in Palo Alto and Pocahontas counties.

Deposits of peat, another natural resource tied to the

The Freda Haffner Kettlehole State Preserve is located along the margin of the Little Sioux River valley (background) in Dickinson County. This steep-sided, bowl-shaped enclosure was formed when an isolated, partially buried pocket of clean glacial ice melted slowly in place.

Mark Engler

state's glacial history, often accumulated in the smaller pond, slough, and bog environments where plant material could not completely decompose. Peat deposits also occur in the form of fens, one of the state's more unusual wetland types. At these isolated hillslope sites, groundwater seeping to the land surface has maintained permanently saturated deposits of peat for thousands of years. The spring-fed habitats support a unique wetland biota, including a number of the state's rare plants. Fens on the Des Moines Lobe are clustered in the hummocky,

end-moraine topography in the northwestern portion of the region. Silver Lake Fen State Preserve in Dickinson County is an outstanding example of these unusual geologic and hydrologic phenomena.

Though good for wetland habitat, incomplete surface drainage is a serious impediment to agricultural productivity. Many of the region's native wetlands were drained as agriculture became more important. Today, numerous low spots contain poorly drained, dark-colored soils indicating sites of previously ponded water and

Douglas C. Harr

A cluster of poorly drained hollows on the Altamont end moraine in Dickinson County supports *Spring Run State Wildlife Management Area*. These recently glaciated landscapes provide important wetland habitats for migrating and nesting waterfowl and other aquatic life.

abundant organic accumulations. Glacial erratics, those travel-worn cobbles and boulders from out of state, are other impediments to agriculture on the Lobe. These prominent relics of glacial activity are seen strewn high and low across the landscape or purposefully gathered into field corners and along fencerows by farmers. Clearing fields of erratics and laying tile lines beneath poorly drained areas have turned the Des Moines Lobe into highly productive farmland. Satellite images of Iowa taken from space, especially in the spring of the year, show a sharp contrast between the intensive cultivation of the productive wetland and prairie soils on the Des Moines Lobe and the different land-use patterns on the more varied soil types of surrounding landform regions (see satellite image, p. 11).

All of the glacial deposits that accumulated from the Des Moines Lobe surges and melting are referred to as the Dows Formation, named for exposures in a quarry near that town on the border of Franklin and Wright counties. Buried beneath the 45 to 60 feet of this glacial drift are several older deposits, including Wisconsinan loess, drift from an earlier episode of Wisconsinan glaciation (the Sheldon Creek Formation, exposed at the surface over part of the Northwest Iowa Plains), various buried paleosols, older Pre-Illinoian glacial deposits, and eventually Cretaceous and Paleozoic bedrock. Sand-and-gravel deposits that occur along the stream valleys mentioned earlier and within the various glacial drifts are potential sources of shallow groundwater for wells in this region. Deeper supplies are available from sandstone or limestone aquifers in the bedrock below.

The grain size and the clay and carbonate mineralogy of the Dows Formation are related to the geographic sources of its ice-transported deposits. As the Des Moines Lobe moved southeasterly across southern Canada, the Dakotas, and Minnesota, it passed over large areas of fine-grained Cretaceous bedrock, especially clay-rich shales and mudstones. Abundant fragments of Cretaceous shale are conspicuous in the Des Moines Lobe glacial deposits. In the field, this obvious characteristic helps distinguish this till from the older Pre-Illinoian tills that underlie the Des Moines Lobe and form the uppermost till deposits in most other regions of the state.

In summary, the uniqueness of this landform region results from its geologically recent encounter with a surging outbreak of ice along the shifting margins of a massive continental glacier. Because this final appearance of glaciers in Iowa took place during and after the period of greatest loess deposition (see next section), the Des Moines Lobe lacks a cover of windblown silt. The result is an exceptionally clear picture of the land surface nearly as the ice left it. In the 11,000 to 12,000 years that have passed since the ice melted, weathering and erosion have made some progress in modifying the landscape. However, the landforms of the Des Moines Lobe still retain the distinct imprints of recent glacial occupation.

Loess
Hills

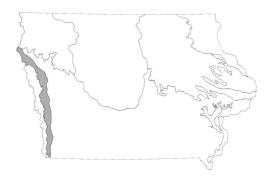

. . . whether viewed . . . in the fall
when they are tinted a rusty red by
the covering of dry blue-joint grass,
or in winter when . . . they present the
aspect of a series of huge snow
drifts, they are of unusual interest
to the physiographer, the geologist,
and the botanist, and they will
some day be more fully
appreciated . . .

—Bohumil Shimek
"Geology of Harrison and
Monona Counties, Iowa," 1910

Irregularities in the landscape always catch the eye. These differences in elevation are described as the land's relief, a trait that accounts for much of a visitor's first impression of a new place. For visitors to western Iowa's Loess Hills, the scenic relief makes a lasting impression.

The irregular Loess Hills form one of the state's most distinctive landscapes. They extend in a narrow band that borders the full length of the Missouri River valley in western Iowa. In the heart of the deep-loess landscapes, usually within two to ten miles of the Missouri Valley, the topography is sharp-featured, with alternating peaks and saddles that dip and climb along narrow, crooked ridge crests (photo, p. 50). Numerous shorter sidespurs branch off the main ridges, and the steeper slopes of both are often horizontally scored with a series of stair-like steps. A dense network of drainageways forming closed-in hollows, narrow ravines, and steep-sided gullies contributes to the intricately carved terrain.

The western boundary of this region is very abrupt—as distinct and well defined as a coastline. The bluffs of steeply pitched, prairie-covered ridges and wooded back-slopes stand boldly apart from the lower, flat-lying cultivated fields of the Missouri River valley floor. Some of the most scenic vistas in Iowa appear from ridgetop summits that overlook the boundary between these two contrasting regions. The eastern boundary of the Loess Hills is not easily defined, as the hills merge gradually with the more rolling landscapes of the Southern Iowa Drift Plain. Three state parks, Stone, Preparation Canyon, and Waubonsie, offer good access to the typical landscapes of the Loess Hills.

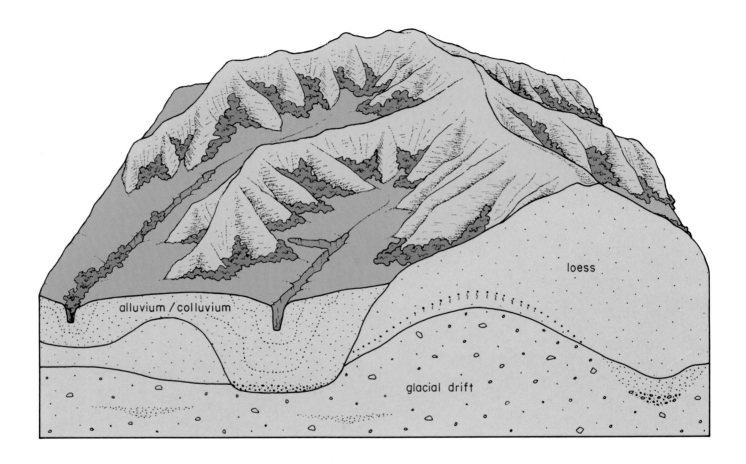

The hills are composed of a gritty yellowish to grayish tan sediment known as loess. Loess is a wind-deposited silt (often including some sand and clay) composed predominantly of closely packed grains of quartz. These deposits in western Iowa are porous, lightweight, and easily eroded. They are vulnerable to collapse when wet yet are quite cohesive when dry, and they exhibit sheer, nearly vertical faces that tend to form slabs, columns, and

The crinkled topography of the Loess Hills features narrow ridge crests, branching sidespurs, steep slopes, and a dense drainage network. The entrance of the Little Sioux River (background) into the Missouri Valley interrupts the sweeping bluffline of the hills in northern Monona County.

vertical tunnels or pipes as they erode. As unusual as these hills appear, loess itself is not a rare or unusual material. In fact, tens of thousands of square miles of the midwestern United States are covered with it. Loess is the parent material throughout broad areas of the nation's most productive agricultural soils. Extensive loess deposits occur along the lower Mississippi Valley, along the Platte River valley of the Great Plains, and in the Palouse district of eastern Washington as well as in central Europe, the Soviet Union, and northern China. Many of these deposits are associated with major river-valley sources, as they are in western Iowa.

The word "loess," usually pronounced to rhyme with "fuss," is German in origin and was first used along the valley of the Rhine River to describe the same types of massive silt deposits. In the early geologic literature of the western Iowa counties, loess was referred to as "silicious marl" or the "Bluff Deposit," and the silt was thought to have originated as a water-laid sediment in a large lake. It was Bohumil Shimek, a well-known nineteenth-century Iowa naturalist studying the fossil remains of air-breathing snails found in Harrison and Monona counties' hills, who first correctly recognized loess as a wind-deposited material.

The deposition of loess was a widespread side effect of Pleistocene glacial activity. Abundant silt and clay in the bedrock along the routes of glacial advance as well as the rigors of glacial transport produced the fine-grained particles. During ice melting, the pulverized debris was released into meltwater streams that drained the ice fronts. During late-glacial time in particular, from about

31,000 to 12,500 years ago, the Missouri Valley served as a major channelway for large volumes of meltwater and sediment released from the vast Wisconsinan ice sheet draped across north-central United States. The region of deep loess in Iowa coincides closely with the wide north-south segment of the Missouri River valley, roughly from Sioux City to Kansas City. It was from this broad valley that windborne silts were swept aloft from floodplains clogged with outwash brought south by seasonally swollen floods of glacial meltwater.

At this time, the Missouri River itself was braided, a tangle of shallow interconnected and shifting channels. During thousands of Pleistocene winters, receding meltwater floods exposed extensive reaches of barren floodplain. Strong, persistent winds, intensified by contrasts in atmospheric conditions between ice-covered and ice-free areas, winnowed the dried floodwater deposits and lifted the finer particles into great clouds of dust. These silt-laden winds blew unhindered across the valley until reaching the rise of the eastern valley margin; here the air currents were deflected, and turbulence caused much of the silty load to drop out. The deepest and coarsest accumulations of loess developed nearest their Missouri Valley source, while smaller amounts of finer particles were carried for miles downwind. The result is a wedge-shaped deposit that becomes thinner into central Iowa. From there, the loess mantle thickens toward the Mississippi as eastern Iowa valley sources added their local contributions to leeward uplands.

The process of loess accumulation took place in varying degrees along all major midwestern valleys that carried glacial meltwater. In western Iowa, however, a combination of abundant fine-grained outwash material, broad valley width, and climatic conditions allowed unusually thick deposits of loess to accumulate. Loess deposition gradually ceased as the discharge of glacial outwash into Iowa's rivers slowed.

While loess deposits are thickest in Iowa and northwestern Missouri, thinner deposits occur on the opposite side of the valley in Kansas, Nebraska, and South Dakota as a result of natural variations in wind direction. The deposits in Iowa, however, are thick enough to obscure the older relief of the preexisting land surface and give their own form to the modern landscape. The resulting terrain is rare among the world's landscapes. Only in the corresponding latitudes of China, along the valley of the Yellow River where massive deposits of loess occur, are thousands of square miles of similar landscapes known to exist. The process of loess deposition is seen today in Alaska along such rivers as the Copper and Matanuska which drain large valley glaciers in the Chugach Mountains along the south-central coast. Though such conditions no longer exist in Iowa, the dry Dust Bowl years of the 1930s or the yellowish brown haze filtering the sun on a windy day during spring planting are reminders of the continuing ability of the wind to sort and move the earth's grit.

The thickness of loess in the Loess Hills is generally over 60 feet, and depths of 150 to 200 feet have been recorded locally from water-well drilling records and outcrop study. It is difficult to say how much the present appearance of the dune-like, corrugated landscape resulted

The slopes of this loess ridge are crossed with a series of stair-like features known as catsteps. Native prairie occupies the drier, exposed summits and sideslopes, while trees and shrubs grow along the more moist, protected ravines and backslopes of this Plymouth County landscape.

Photo by Gary Hightshoe

from the action of wind during deposition. It is clear, however, that many streams within the hills remained active during loess deposition and that much of the easily eroded material has been reworked or removed altogether in the last several thousand years. Erosion has been a major factor in shaping the loess deposits. The intricate, finely sculpted Loess Hills topography is a product of the combined effects of wind deposition, erosional processes along entrenched stream systems, and gravity-induced slumping of thick, fine-grained sediment.

The high-relief terrain within this region has tended to isolate its distinctive landforms from encroachment by agriculture and urbanization and thus protect sizable tracts of its original prairie (see photo, p. xiv). The hills harbor a rich mosaic of specialized ecological niches because of the strong local contrasts in soil moisture, temperature, evaporation, slope angle, and exposure. The dry ridgetops and exposed southwest-facing slopes sustain distinctive shortgrass prairie communities including some unusual native plants and animals. The spine-leafed yucca and plains spadefoot toad, for example, are two species more typically found in desert environments

farther southwest. In contrast, the more protected back-slopes, lower hillslopes, and deep drainageways offer cooler, moister habitats favorable to the growth of trees and shrubs (photo, p. 53). This woodland cover has expanded rapidly over the last hundred years. Prairie fires, important to the maintenance of grassland ecosystems, were suppressed as the region became settled. These timberlands are now so extensive that they are the basis for the new Loess Hills Pioneer State Forest.

The strong connection between landforms, geological processes, and ecology in western Iowa has merited national recognition. Two separate tracts within the Loess Hills have been designated as a National Natural Landmark. This program, administered by the U.S. Department of the Interior, National Park Service, selected nearly 10,000 acres in Monona and Harrison counties as nationally significant examples of landscapes and habitats dominated by loess. The two sites in the heart of this deep-loess country include 7,440 acres north of Turin and 2,980 acres north of Little Sioux.

Three major loess deposits may be distinguished in the Loess Hills. The oldest (at the bottom) is the Loveland, which accumulated throughout the midcontinent during Illinoian time between approximately 120,000 and 150,000 years ago. Next in sequence are the Pisgah (25,000 to 31,000 years old) and Peoria (12,500 to 25,000 years old), which correspond to younger, Wisconsinan intervals of deposition. Each of these wedge-shaped loess deposits is successively younger and more widely distributed from its Missouri Valley source area.

The Peoria loess is the thickest and most commonly seen unit in Iowa. Within this deposit are wind-shaped ripples, clues that the loess accumulated very rapidly at times. Other exposures contain thin dark bands of concentrated organic carbon, marking periods of time when little or no loess deposition took place and the land surface was exposed long enough for vegetation to gain a hold and for soil development to begin. Abundant fossil mollusks, most commonly the fragile white shells of terrestrial snails that lived on the land surface during loess deposition, are also found.

Other color and textural variations appear in loess as a result of alterations caused by weathering. Exposures sometimes show a strongly mottled or splotchy appearance caused by reddish brown iron stains; occasionally, vertical tube-shaped masses of iron called pipestems are also found. These features owe their origins to chemical responses resulting from fluctuations in the water table. Such changes in groundwater-saturated conditions prompted the segregation of iron into these color mottles or into the cylindrical deposits formed along channels of plant roots that once penetrated the loess.

Other interesting objects found within loess deposits are hard nodular pebbles known as loess kindchen (loess dolls). Composed of lime and usually rounded or elongated in shape, these nodules are most often an inch or less in diameter and are usually concentrated in zones. Occasionally they reach baseball or grapefruit size. They were formed by infiltrating rainwater and snowmelt that dissolved and leached carbonate grains present in the loess; as the water moved downward, the lime was concentrated and redeposited around some nucleus to form these unusually shaped concretions.

While the extraordinary thickness of loess is often emphasized, an observant visitor to this region will soon realize that other geologic materials are also present. Older deposits are exposed to view because the loess varies in thickness and in places has been removed by erosion. For example, deposits of glacial drift, including beds of sand and gravel, are frequently seen within the Loess Hills. These older deposits occur beneath the loess and accumulated during Pre-Illinoian glacial events. Abrupt changes observed along hillside profiles, from steep upper slopes to more gentle lower slopes, often coincide with these contacts between loess and underlying glacial till or sand and gravel.

The sand-and-gravel deposits, such as those exposed beneath the loess at Turin in Monona County, hold great interest for paleontologists because of the outstanding fossil remains found in them. Calvin and Shimek's description of the so-called Aftonian fauna was the first documentation of the fascinating array of large and small vertebrates that provide important glimpses of glacial-age environments and animals in Iowa.

Perhaps one of the most unusual geologic materials encountered beneath the loess is another windblown deposit—volcanic ash. A long-studied deposit near the Harrison-Monona county line was dated at 710,000 years old. On the basis of its chemical composition, this 15-inch layer of whitish ash was determined to have originated from eruptions of now-extinct volcanoes in Yellowstone National Park. This and other western Iowa ash deposits are important keys to unlocking the ages and subdivisions of the state's Pre-Illinoian glacial deposits as will be seen in the next section.

Bedrock is another geologic material observed in the Loess Hills, most often in quarries along the bluffline of the Missouri Valley. In fact, the overall width of this valley is a reflection of contrasting bedrock units. The very broad valley along the northern third of the Loess Hills region is underlain by Cretaceous-age rocks consisting primarily of easily eroded shales and thin-bedded, chalky limestones containing abundant fossil clams. These bedrock materials occur quite close to the land surface in the northern Loess Hills and are easily seen in the Sioux City area. From Harrison County south, however, the region is underlain by Pennsylvanian limestones, and the Missouri Valley narrows abruptly in response to this more resistant bedrock type. The presence or absence of these various geologic materials near the land surface, combined with their uneven resistance to the lateral erosion of the Missouri River, helps explain the sweeping, crescent-shaped embayments that indent the valley's bluffline.

Erosion in the Loess Hills deserves special mention. Much of this landscape's unique appearance results from erosional activity coupled with the special physical properties of loess mentioned earlier. For example, erosion at the base of the bluffs along the Missouri Valley has taken place primarily through contact with flowing water. The Missouri River along much of its length in Iowa flows several miles west of the Loess Hills; however, this river or one of its tributaries clearly has swung eastward against the bluffs, gnawed at their base, carried away

Loess is highly erodible and unstable when wet, which presents land-use hazards in the region. Deep, narrow gullies, which can lengthen and widen quickly after rainstorms, are characteristic erosional features at the upper ends of smaller drainageways.

the collapsed loess, and left sheer slopes and an abrupt bluffline—perhaps only within the last few hundred years. In addition, ridges of loess once extended out into the valley, perpendicular to the present bluffline. These have been planed off, leaving steep, triangular-shaped faces known as truncated spurs. Older, higher floodplain levels (terraces) within the Missouri Valley also have been removed by erosion, though major tributaries such as the Soldier, Maple, and Boyer rivers still shelter these landform features and their geologic records. The principal landscape features formed along the Missouri Valley

margins since this erosional activity are alluvial fans. These apron-shaped deposits flare outward from smaller sediment-laden tributary streams that spill from the hills. On steep upper slopes within the hills, erosion also takes place by dislodgement of loess, encouraged by the pull of gravity. Catsteps, the striking stair-like features that mark many of the steeper slopes, result from periodic slipping and irregular downslope transfer of loess by gravity. These small slumps, good indicators of unstable slopes, also may form in response to trampling by livestock.

The physical properties of loess also contribute to a number of engineering problems and land-use hazards. For example, the style of roadcuts throughout the region reflects the capacity of exposed loess to maintain nearly vertical faces. Such excavations usually have steep single walls or are stepped back in a series of steep risers and horizontal treads. Cracks tend to develop along natural vertical partings or zones of weakness that often occur behind and parallel to these exposed faces of loess. Porosity and the natural steepness of loess slopes promote their relatively dry condition and apparent cohesiveness. During wet periods, however, when infiltrating water lubricates these vertical openings or when a rising water table saturates the base of an exposure, the seemingly stable loess can no longer bear its own weight. It collapses easily, sometimes resulting in serious landslides. Fresh scars on the landscape where these slope failures take place are prominent after heavy rains, and road-maintenance crews must periodically reopen roads blocked by slumped loess. The relocation of the War Eagle Monument on the bluffs at Sioux City was necessitated by recurring episodes of slope instability and collapse.

The erodibility of loess and its instability when wet pose other serious problems and land-use hazards in this region. Soil erosion rates are very high, and the amount of eroded sediment carried in streams draining the region is among the highest recorded in the United States. Gully erosion is especially pronounced, and these deep, narrow, steep-sided features are characteristic of the region's smaller drainages (photo, left). Gullies lengthen headward, deepen, and widen quickly after rainstorms, cutting into cropland, clogging stream channels and drainage ditches, and forcing costly relocations of bridges and pipelines.

The Loess Hills are undoubtedly Iowa's most fragile landform region in terms of susceptibility to erosion. The land is being shaped as rapidly as any terrain in the state. It is also clear from the geologic record that these dynamic episodes of erosion and deposition have been part of the scene for thousands of years. While gully erosion is intensified by stream channelization, loss of vegetation cover, overgrazing, cultivation, and other types of human disturbances, recent geological studies have shown that episodes of gully cutting and filling have occurred repeatedly during the last 25,000 years, significantly rearranging some of the thick accumulations of loess. Studies of buried gully fillings have also revealed previously unsuspected archaeological remains preserved because of these geological processes. The search for other archaeological sites can be guided by an improved understanding of the region's landscape evolution.

The emphasis on erosion and the realization that much of its impact on the landscape is geologically quite young are important themes that will be repeated in the following sections on Iowa's landform regions.

Southern Iowa Drift Plain

An east-west traveler must cross a series of alternating ridges and valleys. The north-south traveler may usually find a ridge road. From the latter, looking off over the country, the tops of the successive flat-topped ridges appear rising to an even surface and restoring the old plain in which the valleys have been carved.

—H. Foster Bain
"Geology of Decatur County,"
1898

The topography of the Southern Iowa Drift Plain is perhaps most representative of "typical" Iowa landscapes. The Southern Iowa Drift Plain is certainly the largest of Iowa's landform regions, and it is the primary region travelers on Interstate 80 will see. Noted Iowa artist Grant Wood emphasized the steeply rolling character of this landscape in many of his stylized paintings, especially *Young Corn* and *Fall Plowing*. The landforms of this region, like those of the *Des Moines Lobe*, are composed primarily of glacial drift, but the massive ice sheets that carried material into this part of Iowa were older by hundreds of thousands of years than those that occupied the north-central area of the state. It is this geological generation gap that makes all the difference in the appearance between these two regions.

Features typical of a freshly glaciated landscape have been obliterated by time. Gone are the moraines, kames, kettles, bogs, and lakes—all those distinctive visual clues to recent contact with glacial ice. The only remaining evidence to verify passage of these earlier ice sheets is the ten to hundreds of feet of glacial drift covering the bedrock surface. Instead of poorly drained, low-relief landscapes, streams have had time to establish well-connected drainage systems and to carve deeply into the land surface. Hillslopes, especially those higher in the drainage network, often display a texture of finely etched rills which give a distinct ribbed or furrowed appearance to the terrain. These rills give way to ravines, then to creeks that flow part of the year, and eventually to perennial streams and rivers in major valleys. Patterned like the branching veins in a leaf, this dendritic network has

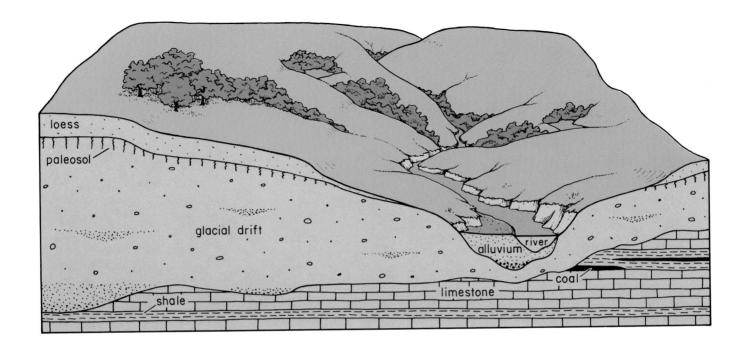

loess
paleosol
glacial drift
alluvium
river
shale
limestone
coal

drained the postglacial wetlands, erased the ice-contact landforms, and through time has reshaped these old glacial plains into the deeply creased landscapes so familiar in this region today (photo, p. 60).

With most of the land surface sloping toward some drainageway, the terrain of this region projects a feeling of enclosure when we travel among its hills. Views extend only as far as the next rise or the next bend in the road. There are no long-distance vistas except those seen from hillcrests which return again and again to the same elevation, each time providing a glimpse over the billowy landscape beyond. At night, farmstead lights suddenly dot the darkness, or approaching headlights abruptly drop from sight as the road alternately rises and falls.

The observation that Southern Iowa Drift Plain summits always seem to return to a uniform elevation is a clue to their geologic origins. These even-topped uplands disclose the approximate level of the original, once-con-

tinuous land surface constructed by the last ice sheet to pass this way. Every hillslope and valley floor mark the extent of erosion into the old glacial plain. The space between hills emphasizes the great amount of material that has been removed and the hundreds of thousands of years this process has taken.

The erosional processes that have carved these hills from the earlier glacial plain have not been at work continuously or uniformly through time. Instead, the past has been punctuated with episodes of rapid erosion accompanied by valley deepening and lengthening. The episodes of downcutting alternated with periods of greater landscape stability when soil profiles could weather deeply into the exposed glacial deposits. This variable intensity in the erosional shaping of the landscape actually left broad steps notched into the hillslopes of the region's drainage basins. These stepped erosion surfaces occur throughout the portions of Iowa where Illinoian and Pre-Illinoian glacial deposits are the dominant landscape materials. Thus, hillsides are a key element of Southern Iowa Drift Plain landscapes. They may appear at first glance to be smoothly flowing slopes, but subtle changes

Old glacial plains have been reshaped by erosion into steeply rolling hills and valleys typical of the Southern Iowa Drift Plain. Near the Iowa-Poweshiek county line, strips of row crops alternating with hayfields emphasize changes in hillslope gradient—a key to variations in the region's erosional history.

Photo by Gary Hightshoe

of contour from more steep to less steep reveal past irregularities during their long erosional history.

As this dissected landscape was evolving, a windblown mantle of loess was added to the land surface. This silty (and sometimes sandy) deposit usually ranges in thickness between 5 and 30 feet throughout the region and is deep enough in places to add to the local relief, particularly on leeward hillsides and along valley margins. The thickest loess deposits over the region are found close to major sources of windblown silt. Thick deposits also are preserved on broad, uneroded uplands where the most continuous accumulation occurred, while on narrower divides and hillslopes loess deposits generally are much thinner because of erosion. The bulk of this silt mantle is composed of Peoria Loess. Beneath this loess in the western part of the region is an additional thin wedge of the older Pisgah Formation loess, and below that the still older, widespread Loveland Loess occurs. These windblown deposits are thickest near the Missouri River valley and become thinner toward the central portion of the Southern Iowa Drift Plain. The Peoria Loess then thickens again nearer the Mississippi River and other eastern Iowa valley sources. As seen in the previous section, the influence of this windblown silt in extreme western Iowa dominates the appearance of an entire landform region.

Across the loess-mantled Southern Iowa Drift Plain, remnants of four major landscape levels are preserved. Some of these are prominent, platform-like features, while others are seen only as a gentle flattening of the gradient along a hillslope profile. The four landscape lev-

61

els, or surfaces, were progressively eroded into the old glacial plain, leaving the oldest landscape surface at the highest elevation; the newer, younger surfaces each cut into lower landscape positions and into stratigraphically older material. The four surfaces (highest to lowest and oldest to youngest) are known as the Yarmouth-Sangamon, the Late-Sangamon, the Wisconsinan (Iowan Surface), and the Holocene (postglacial). These surfaces differ in age and in the thickness of their loess mantle. They also vary in their preservation and dominance in the landscape from one place to another. It is helpful to keep in mind that most of these Pleistocene surfaces originally developed on landscapes that are no longer part of the present land surface because they were buried by younger deposits or were pruned back by later episodes of erosion. As a result, today's hillslopes are not underlain by uniform materials of the same age; instead, descending slopes cut across a series of paleo-landscapes and a sequence of different glacial-age strata. Because these Pleistocene materials are relatively soft, at least compared to bedrock, they leave smooth shapes in the landscape. There are few obvious, abrupt breaks in slope to highlight where these changing stratigraphic relationships occur; there are, however, some useful clues.

The highest and oldest landscapes in the region are flattopped upland summits. These broad, loess-mantled uplands are largely uneroded remnants of the last Pre-Illinoian drift plain that lay exposed to weathering during the Yarmouth, Illinoian, and Sangamon stages. This Yarmouth-Sangamon surface has a distinct loess-mantled paleosol. Where it outcrops on hillsides beneath the loess, the paleosol is usually noticed as a thick, gummy gray clay. This ancient soil profile, once referred to as "gumbotil," often has a pronounced effect on local drainage conditions because it acts as a barrier which retards the downward movement of groundwater. Rainfall and snowmelt percolating through the loess tend to move laterally once they reach the less permeable clay of this paleosol. Seeps or springs often develop on hillsides where the clay-rich zone is intercepted by the land surface. Such areas of "gumbo," notoriously sticky when wet and rock-hard when dry, are well known to farmers working the fields of this landform region.

Throughout the Southern Iowa Drift Plain, a gradual topographic shift occurs in the position and amount of level terrain from east to west. It is obvious in parts of southeast Iowa that extensive areas of nearly flat, uneroded uplands—the Yarmouth-Sangamon surface—are more common than the typical steeply rolling hills. In many locations, around Ottumwa, Fairfield, Mount Union, or Mediapolis, for example, these tablelands (tabular divides) are the strongest visual element of the landscape, and steep, hilly, wooded terrain occurs only near stream valleys. The amount of level land along valley floodplains is small in comparison with that on these extensive upland areas.

The presence of these undissected uplands in southeast Iowa seems to reflect the influence of shallow bedrock. Erosion of the glacial deposits, very common elsewhere in the region, has been slowed here because streams draining these uplands encounter more resistant Mississippian limestones and dolomites in their valleys

Drake Hokanson

With most of the land surface in slope, east-central Iowa (Jones County shown here) often appears as a richly diverse mosaic of productive cropland, pasture, groves of timber, and sturdy reflections of its long rural heritage.

and thus are inhibited from cutting deeper and farther back into the upland divides. Karst features, such as caves, springs, and sinkholes, which are abundant in northeast Iowa, also are seen in this region in the Burlington area where limestone occurs particularly close to the land surface.

Below the high upland summits of the region, the younger Late-Sangamon surface is frequently seen as a gentle shelving or flattening of the gradient along hillslopes. From beneath its loess mantle the thinner but distinct paleosol that marks this former land surface appears as a rust-colored zone caused by oxidation of iron particles. This clayey paleosol (see photo, p. 23) is commonly seen streaking freshly plowed hillsides, particularly in the spring and fall when soil color is most easily noticed.

The terrain of south-central and east-central Iowa is more dissected, more deeply cut by streams. The nearly level upland divides, so prominent in southeastern Iowa, are much smaller in area, and most of the land surface is in hillslope; sightings of the Late-Sangamon paleosol are common. This topography (photo, left) is typical of an area that includes Monticello, Iowa City, Oskaloosa, Centerville, Chariton, and Indianola.

A still younger Wisconsinan erosional step can be observed crossing lower hillsides within the Southern

63

Iowa Drift Plain. This narrow, subtle surface is distinguished by a very thin loess cover and the noticeable absence of any paleosol. The Wisconsinan erosion surface was cut into the landscape while loess was being deposited during the intense glacial cold that gripped the midcontinent between 16,500 and 21,000 years ago. The best topographic expression of this erosion surface is seen throughout the region mapped as the Iowan Surface, where it extends across virtually the entire landscape. In the Southern Iowa Drift Plain, however, its gentle incline occupies just the lower slopes of drainage basins. An interesting stratigraphic feature sometimes seen on both the Late-Sangamon surface and this younger Wisconsinan surface is a stone line or pebble band. This residual or lag deposit of glacial gravels was concentrated at the land surface over a period of time as erosion removed the surrounding finer materials.

The topographic shift across the region continues into southwestern and western Iowa where the flat upland summits disappear almost entirely. The hills here appear aligned in long, parallel crests of steep waves with broad troughs between them, as seen in the areas around Denison, Atlantic, Red Oak, and Creston. The most extensive areas of level terrain occur along valley floors. The floodplains here and elsewhere across the state thus mark the lowest and youngest (Holocene) erosion surface cut into the landscape. This surface is marked by postglacial alluvial deposits.

Throughout the Southern Iowa Drift Plain, the relative amounts of level upland divides, steeply rolling hillslopes, and lowland valley floors vary considerably. However, the arrangement of these various landscape elements, the relief resulting from the combined episodes of erosional history, and the exposed paleosols that sometimes color the hillslopes are the dominant unifying characteristics of the region.

Investigations since the mid-1970s have substantially revised the stratigraphic subdivisions of the older glacial deposits that form the bulk of this region's landscapes. Traditional glacial and interglacial terms such as "Kansan," "Aftonian," and "Nebraskan" have been abandoned. The concept of one till per glacial period, as interpreted from exposures of glacial deposits, did not hold together after core-drilling at many of the classic localities in western Iowa revealed as much as 200 feet of additional glacial deposits below some of the previously studied outcropping units. Geologists then recognized that the glacial sequence in Iowa is quite complicated, with more glaciations, at least seven more till sheets, and more buried soil profiles than were previously described. As a result, all Quaternary deposits older than Illinoian (greater than 300,000 years old) are, for now, referred to collectively as Pre-Illinoian.

The work in western Iowa was followed by detailed landscape analysis, outcrop descriptions, and drilling investigations over large areas of eastern Iowa. The fieldwork was combined with laboratory studies of various mineral grains within the Pre-Illinoian till deposits, as well as a reevaluation of where and how many paleosols occurred. The various physical and mineralogical characteristics that emerged from these studies permit the division of Pre-Illinoian glacial deposits into distinct lithologic units named the Wolf Creek Formation and the older Alburnett Formation.

In addition to Pre-Illinoian glacial deposits, the Southern Iowa Drift Plain also includes the small area of eastern Iowa covered by Illinoian glaciation about 300,000 years ago. Only the oldest of several Illinoian glacial advances actually entered Iowa. This advance was part of an ice mass known as the Lake Michigan Lobe, which covered most of Illinois as it extended southwesterly from a larger continental ice sheet. The deposits that remain in Iowa from this eastern-source glacier are known as the Glasford Formation (Kellerville Till Member). They contain an abundance of Pennsylvanian rock fragments, including coal and plant fossils common to the bedrock of eastern and central Illinois. These materials contrast sharply with Iowa's Pre-Illinoian till deposits, left by glaciers which crossed much different Tertiary, Cretaceous, and Precambrian rock types as they advanced along flow paths from the north through the Dakotas, Minnesota, and the Lake Superior area before pushing completely through Iowa into Missouri. In this area of Illinoian glaciation, however, as with the region as a whole, sufficient time has elapsed for erosion to reshape the glaciated terrain and establish an integrated drainage network, producing landscapes that resemble the rest of the Southern Iowa Drift Plain.

Further supporting the major revisions of Iowa's early glacial record is the fascinating evidence provided by volcanic-ash deposits found sandwiched between some of the Pre-Illinoian glacial tills in western Iowa. These gritty, whitish gray deposits are windblown particles consisting of glass shards and pumice fragments. Their chemical and mineralogical composition can be traced to eruptions of now-extinct volcanoes in Yellowstone National Park in northwestern Wyoming, approximately 850 air miles west of Iowa. Preserved remnants of ash falls from these eruptions are known to occur in Woodbury, Guthrie, Adair, Union, Ringgold, Harrison, Monona, Audubon, and Cherokee counties. No deposits of volcanic ash have been found in the eastern part of the state.

Through a technique known as fission-track dating, four separate ash falls from the Yellowstone area have been dated in Iowa and Nebraska. They have ages of 0.6, 0.7, 1.2 and 2.2 million years and as a group are referred to as the Pearlette Family of volcanic ash deposits. The 2.2-million-year-old ash overlies the oldest of at least seven separate Pre-Illinoian glacial deposits recognized in Iowa. This pinpoints the beginning of glaciation here as sometime before 2.2 million years ago—considerably older than many previous estimates. This dating is important to establishing a time frame for Iowa's glacial record and correlating glacial-age events here with those in other parts of the world.

The valleys eroded into these old glacial deposits are among the most picturesque features of the Southern Iowa Drift Plain. Many of the larger river valleys, especially the Cedar, Iowa, Skunk, Des Moines, Boyer, and Little Sioux, had glaciers standing in their headwaters during the time the Des Moines Lobe was ice-covered. These valleys obtained much of their present width, depth, and alluvial fill during meltwater flooding as the

Rills and streams branch out across the southern Iowa landscape forming a well-integrated drainage network. Farm ponds and grassed waterways help store water resources and soil moisture.

Wisconsinan ice sheet disappeared from north-central Iowa. In many places the rivers have carved completely through the upland sequence of loess, paleosols, and glacial drift into the layers of sedimentary bedrock beneath.

The rough, wooded terrain adjoining these deeper valleys supports many scenic recreation areas and important wildlife habitats. The dendritic patterns of streams established across the landscape result in mosaics of cropland, pasture, and timber. Contour plowing, strip-cropping, and grass-backed terraces often are used to minimize soil erosion and conserve water on the cultivated hillsides.

In addition, the steeply rolling terrain is suited to building dams to form reservoirs. Three of Iowa's largest reservoirs are in this region—Coralville, Red Rock, and Rathbun. These structures control flooding and also provide recreational opportunities and sources of water. The shorelines of these reservoirs are indented with numerous inlets marking the locations of smaller tributary valleys in the formerly well drained landscape.

Across large areas of the Southern Iowa Drift Plain, groundwater is not as plentiful as in other parts of the state. For example, the thick sequence of shale-dominated Pennsylvanian bedrock underlying central and southwestern Iowa is a notoriously poor source of groundwater for wells. Many rural residents in southern Iowa depend on large-diameter wells that tap shallow groundwater seeping along the contact between the loess and underlying, less permeable glacial till. Farm ponds are a characteristic feature of the steeply rolling landscapes. Trapping surface runoff in these small reservoirs provides an important supplementary source of water for rural Iowans (photo, left).

In the eastern third of the region, limestone suitable for road construction and maintenance as well as agricultural use is quarried from sites underlain by Mississippian, Devonian, and Silurian rocks. Geodes, the state rock, are sought by collectors from streambeds cut into Mississippian shale (Warsaw Formation) along tributary valleys of the Des Moines River in extreme southeastern Iowa.

The Pennsylvanian rock formations in the southwestern two-thirds of the region contain Iowa's historically important coal deposits. Surface mining once was common in many areas where coal seams occur at shallow depths, especially in south-central counties. Recontouring and replanting these abandoned mine lands, as well as identification of areas where cave-ins can occur over abandoned underground mines, are important environmental objectives in the region. Plant fossils also occur in these Pennsylvanian rocks. Collectors can find fossil leaves and segments of branches, bark, and roots of the lush tropical vegetation that flourished in the coastal swamps of Iowa about 410 million years ago.

The Southern Iowa Drift Plain displays an intriguing variety of landscapes. Their shapes result primarily from the deepening network of rivers and streams. Because of the topographic relief resulting from this long-term erosional activity, the earth materials exposed along hillsides reveal more of the state's glacial-age history than is seen in any other region.

Iowan Surface

. . . a frozen sea, flecked with wakes of dark shadow behind each great boulder, fretted here and there with long, low rollers . . . and again broken by higher but shorter swells, gracefully curved as the bending backs of giant dolphins . . .

—W. J. McGee
"The Pleistocene History of Northeastern Iowa," 1891

Sweeping, outstretched landscapes span a large area of the northeastern quarter of Iowa. This relaxed, open topographic style is free of the strong expressions of glacier surges, silt-laden winds, or erosional sculpture that identify the Des Moines Lobe, the Loess Hills, and the Southern Iowa Drift Plain. Part of the individuality of this particular landform region lies in its topographic restraint and subtle land shapes. The tendency not to draw attention to itself also suggests that the region's landscape features and deposits do not offer an easily recognized geological explanation for their appearance. In fact, the region was a focus of much scientific controversy until new research methods and the concept of stepped erosion surfaces were used in the 1960s to challenge the generally accepted existence of a separate sheet of Wisconsinan glacial drift in this area.

The physical characteristics of this landform region, identified now as the Iowan Surface, contain numerous clues to an elusive chapter of Iowa's glacial history. The land surface usually appears slightly inclined to gently rolling with long slopes, low relief, and open views to the horizon. This contrasts sharply with the restricted lines of sight noted within the more billowy, steeply rolling landscapes of the Southern Iowa Drift Plain. Like southern Iowa, however, the hillslopes of the Iowan Surface can be described as having multi-leveled or stepped surfaces. These levels, though subdued, occur in a gradual progression from the major stream valleys outward toward the low crests that mark their drainage divides. It is difficult to pick out a clearly defined valley edge; more often the eye sees only a series of long slopes merging almost

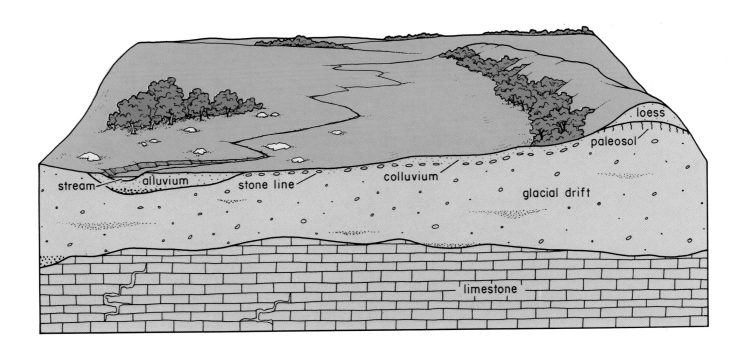

Labels on figure: loess, paleosol, stream, alluvium, stone line, colluvium, glacial drift, limestone

imperceptibly with a gentle rise to the next watershed divide. Drainage networks are well established, though stream gradients usually are low and some scattered areas of poor drainage and original wetlands occur. The low-relief landscapes offer very few exposures of their internal composition.

Erosion on a large scale is the key to the geological origins of the Iowan Surface. Earlier in its landscape evolution, before the Wisconsinan glacial events, the region was actually part of the Southern Iowa Drift Plain. It was last visited by glaciers in Pre-Illinoian time and since then has lain exposed to various episodes of weathering and soil development, erosion, and loess deposition. We have already seen that past erosional activity in Iowa has not been uniform through time and that episodes of more accelerated erosion have cut broad steps or levels into the slopes of the state's watersheds. One of the latest of these episodes occurred between 16,500 and 21,000 years ago during the coldest part of the Wisconsinan, just as the continental ice sheet was becoming fully expanded.

69

In the area of the Iowan Surface, the intensity of this cold-climate weathering and erosion simply overwhelmed the earlier landscape contours. Earth materials were loosened, removed, and redeposited across virtually the entire landscape, leaving only scattered uneroded remnants of higher ground.

Fossil remains of arctic and subarctic species of plants, insects (especially beetles), small mammals, and snails have been recovered from organic-rich sediment in eastern Iowa glacial deposits dating from this interval. The fossil assemblage, remarkably close to flora and fauna living in northern Canada today, provides evidence that tundra conditions were well established in Iowa 17,000 years ago. During this time climatic conditions were substantially colder and wetter than they are at present. Extensive freeze-thaw action, massive dislodgement of loosened material, sheetwash of slopes, and turbulent winds were forms of erosional scouring that took place throughout the cold but ice-free tundra-covered areas of the upper Midwest. The effect of these periglacial climatic conditions and geologic processes was to vigorously plane the deeply eroded landscapes across northern Iowa, the area in closest proximity to the spreading ice front. Under these conditions, hillslopes became very unstable and were worn down as the land slumped and washed. The Pre-Illinoian upland summits and divides were lowered, and the Yarmouth-Sangamon and Late-Sangamon paleosols were almost completely stripped from the landscape.

The most pronounced topographic effects of this erosional scrubbing of the landscape occurred in the northern half of the state, reflecting the colder, more severe climatic conditions. The effects of this erosional episode, however, also spread into southern Iowa but with less noticeable results. In the Southern Iowa Drift Plain, this erosion surface is seen only as a minor leveling of the landscape along the lower slopes of watersheds. Furthermore, the low-relief Iowan Surface often is noticeably inset below the steeply rolling landscapes of the Southern Iowa Drift Plain, a relationship best observed where the two regional boundaries are in sharp contact. A north-south trip through Benton County in the vicinity of Blairstown, for example, offers a good view of these contrasting landscapes and their relative differences in elevation.

In north-central Iowa, this widespread erosion surface was later overridden by the Des Moines Lobe ice sheet, whose deposits and landforms mark the western border of the Iowan Surface. To the east, the Paleozoic Plateau contains only isolated pockets of any Pre-Illinoian glacial deposits. The border between the Paleozoic Plateau and the Iowan Surface is marked by the Silurian Escarpment, a massive leading edge of Silurian-age strata that reflects a major difference in erodibility between geologic materials on either side of the boundary. The southern border of the Iowan Surface is very irregular, as it is crossed in numerous places by major northwest-to-southeast trending stream valleys. The border is further masked by an increasing thickness of windblown loess and sand, which originated from these valleys as they were filling with glacial meltwater and sediment. So, while the Iowan Surface has a distinctive topography and a mappable boundary, many of its characteristics actually continue to be expressed beyond its borders, though buried beneath younger deposits in the case of the Des Moines Lobe to the west, or taken over by more expressive bedrock materials in the case of the Paleozoic Plateau to the east, or diminished by increasing distance away from the latitudes of greatest climatic intensity in the case of the Southern Iowa Drift Plain. The result is that the Iowan Surface is a regional window through which we view the great reduction in topographic relief that was imposed by erosion on the landscapes and Pre-Illinoian glacial deposits of northern Iowa.

Loess deposition was also under way during this time, but for the most part loess could not accumulate on the rapidly evolving landscape in the face of such corrosive erosional activity. Only toward the end of this episode did a thin, discontinuous deposit of loess manage to stay in place on the land surface, primarily in the southern part of the region. In addition, the more resistant pebbles and cobbles from the glacial drift lagged behind on the eroded slopes. This residual stone line or pebble band is a commonly observed stratigraphic feature where roadcuts or quarries expose cross-sectional views of the eroded drift surface and the thin loess or loamy sediment that may cover it.

Other features typical of the Iowan Surface are the ever-present fieldstones known as glacial erratics. Composed of igneous and metamorphic rock types that form the bedrock foundations of Canada, Minnesota, and Wisconsin, they were part of the great volumes of rock and soil material scooped up by the advancing Pre-Illinoian glaciers and suspended within the frozen masses of moving ice. Eventually released by glacial melting, these travel-worn boulders are now stranded far south of their native areas. Great numbers of glacial erratics are observed across the Iowan Surface, especially along its shallow valleys (photo, p. 72), indicating the boulders' resistant composition withstood the erosional wear and tear that removed so much of the softer, clay-rich deposits that once enclosed them.

Though more recently deposited glacial erratics are also common on the Des Moines Lobe, it is the unusual size of many Iowan Surface boulders that characterizes this region. Many of the largest erratics have been broken apart by dynamite for use in building foundations or simply to remove them from cropped fields. Some examples of the few remaining large erratics can be seen in the city park at Nora Springs, in a farm field along Floyd County route T-70 three miles west of Nashua, along a gravel road one mile south of U.S. 20 on the Grundy–Black Hawk county line five miles west of Cedar Falls, and at St. Peter's Rock three miles southeast of Alta Vista in Chickasaw County. A large granite boulder also is present in Grammer Grove, a wooded Marshall County wildlife area near Liscomb. Well drillers throughout the region occasionally report unexpected encounters with enormous buried erratics which resist drill bits and require moving well sites some distance away.

Smaller erratics continue to work upward to the land surface during seasonal freezes and thaws. They are well known to farmers in the region who often haul them to unused field corners or pile them along fence lines. Pas-

Long, gently inclined slopes with unrestricted views to the horizon are typical of the Iowan Surface. Broad, shallow valleys and abundant glacial boulders, as seen here in southeastern Black Hawk County, reflect a landscape shaped by erosional scour during intense glacial cold.

tures in Chickasaw, Bremer, Butler, Buchanan, and Black Hawk counties are good places to look for undisturbed boulder-strewn landscapes. Glacial erratics, though present, are less commonly seen in the Southern Iowa Drift Plain because of the loess cover and the slower geological progress in eroding the glacial drift that surrounds them. Glacial boulders in that region are usually seen in ravines and valleys where downcutting by streams has exposed them.

In the southern third of the Iowan Surface, slopes steepen near the larger stream valleys. Elongated ridges and isolated oblong hills known as paha occasionally rise prominently (30 to 100 feet) above the surrounding plains (photo, p. 74). The term "paha" comes from Dakota Sioux dialect meaning "hill" or "ridge," and it was first applied in 1891 by W. J. McGee to the special hill forms in this region of Iowa. These streamlined landscape features, mantled with silt and sand, are concentrated along the region's irregular southern border and are distinctly oriented in a northwest-to-southeast direction. Eighty percent of the 116 paha mapped in this region are within 22 to 50 miles of the southern border. Their distribution and alignment parallel to (and often very near) river valleys strongly suggest that paha are actually wind-aligned dunes that accumulated in response to the strong, prevailing northwest winds that were scouring the Iowan Surface during this period of glacial cold. The paha formed and migrated in close proximity to those valleys that supplied abundant amounts of loose sand and silt. Their solitary landscape positions close to the southern border of the region also indicate that, at least locally, rapid accumulations of massive amounts of loess managed to stay ahead of the wasting landscape and protect the older glacial deposits underneath from further erosion.

Paha also occur along interstream divides, which along with the sequence of materials identified in drilled core-samples taken from their interiors further confirms their origins as erosional remnants preserved by thick windblown deposits. Paha are the last topographic and stratigraphic remains of uplands that were once part of a higher, older, continuous Pre-Illinoian glacial plain. Internally, the paha are underlain by Pre-Illinoian tills, many with a well-developed gray (Yarmouth-Sangamon) or reddish (Late-Sangamon) paleosol still intact at the top.

As noted, they are capped with a thick interval of loess and often sand. This is the same stratigraphic sequence normally found throughout the uplands of the Southern Iowa Drift Plain.

Soils, especially those mapped on the larger paha, indicate the subsequent native vegetation of these elevated sites was forest rather than prairie. Perhaps the prairie fires that maintained the surrounding grasslands detoured around these hilltop positions just as they bypassed low-lying wooded river valleys across the state. Also noted on larger paha are differences in permeability between the more porous loess and sand and the underlying clay-rich paleosol and glacial till. These hydrologic conditions produce occasional hillside seeps, just as was noted among southern Iowa's landscapes. Casey's Paha (Hickory Hills Park) in northern Tama County and the campuses of Cornell College in Mount Vernon and Kirkwood Community College in Cedar Rapids are prominent paha landmarks in eastern Iowa and are good examples of McGee's "graceful, dolphin-backed hills."

Only a few scattered paha occur in the northern two-thirds of the Iowan Surface. In fact, the remaining glacial deposits of this area are quite thin, and the influence of shallow limestone bedrock is seen on the land surface in the form of karst features, especially sinkholes. Good examples of sinkholes, where cavities in the underlying Devonian limestone resulted in collapse of the land surface, are seen in farm fields along U.S. 218 north of the town of Floyd in Floyd County. Sinkholes and broad sags in the landscape, where karst conditions are just beginning to show above shallow fractured limestone, make

this intensively cultivated region especially vulnerable to groundwater contamination. As we will see, the effect of underlying bedrock on the landscape is magnified considerably in the Paleozoic Plateau, which is that much closer to the influence of the Mississippi Valley.

Limestone and dolomite deposits in the eastern half of Iowa are abundant, and their high quality and relatively shallow depth support an important economic mineral industry in this part of the state. Rock quarries excavated into Silurian and Devonian formations are especially common within the Iowan Surface where these bedrock units usually are not too deeply buried by glacial materials. The carbonate rocks also provide substantial quantities of groundwater to wells throughout the region. Because of the thin glacial drift, the flow of major rivers such as the Winnebago, Shell Rock, Cedar, and Wapsipinicon is maintained during dry conditions by the contributions of groundwater discharge from these bedrock aquifers to the streambeds.

Groundwater flow also sustains unusual local wetlands known as fens. These water-saturated peat deposits, supporting some of the state's rare plant communities, were described earlier as being clustered in the northwestern part of the Des Moines Lobe. Fens are also present, though much more widely scattered, among the gentle slopes of the Iowan Surface.

In the past, the surface deposits of this region were regarded as glacial drift from a Wisconsinan ice advance into Iowa prior to the Des Moines Lobe. The paha were regarded as possible mounds of till molded by that advance or as high areas bypassed by thin glacial ice. Improved research methods based on core-drilling transects across the landscape proved convincingly that these theories were not true and that the so-called "Iowan drift" does not exist. This region evolved not from glacial deposition but from normal processes of erosion in cold environments where frost action, downslope movement of water-soaked soil materials, and strong winds were the dominant geologic processes at work.

Solitary, elongated hills, known as paha, rise prominently above gently rolling landscapes. This Linn County paha near Mount Vernon, oriented in a northwest to southeast direction, is a remnant of an earlier glacial plain preserved by a cover of windblown silt and sand.

Photo by Timothy J. Kemmis

Northwest Iowa Plains

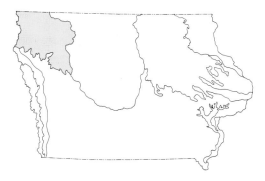

. . . there are gradual slopes from a broad crest leading to a broad valley. To gain one crest, however, is simply to discover another hollow. The grades are not steep, they may usually be climbed with a bicycle, and the long even declines make it possible to coast to the bottom . . . and part way up the next slope.

—Frank A. Wilder
"Geology of Lyon and Sioux Counties," 1900

The landscapes of northwest Iowa have a seasoned appearance, exhibiting a blend of terrain features and geologic materials that individually dominate other landform regions. Taken together, however, they produce a distinctive combination that sets apart this region of the state.

The gently rolling landscapes of the Northwest Iowa Plains are reminiscent of the low, uniform relief seen on the Iowan Surface. A well-established branching network of streams reaches out over all of northwestern Iowa, providing effective drainage and a uniformly creased land surface. Most of the valleys are wide swales that merge gradually with long, even slopes up to broad, gently rounded interstream divides. This uniform density of stream drainage is important in unifying a region where both Pre-Illinoian and Wisconsinan glacial deposits occur at the land surface.

The uplands in the western half of this region are underlain by Pre-Illinoian glacial tills that have been erosionally stripped of their paleosols. In the eastern half of the region, however, these tills are covered by younger glacial deposits. This sheet of glacial drift, mapped as the Sheldon Creek Formation (see map, p. 34), links the Northwest Iowa Plains to the Wisconsinan glacial advances displayed so vividly in the landscapes of the Des Moines Lobe to the east. The Sheldon Creek drift is also Wisconsinan in age but was deposited during an earlier glacial episode. Radiocarbon dates place the age of this older ice advance at about 20,000 to 30,000 years ago or 6,000 to 16,000 years earlier than the Des Moines Lobe advance. Deposits of the Sheldon Creek Formation

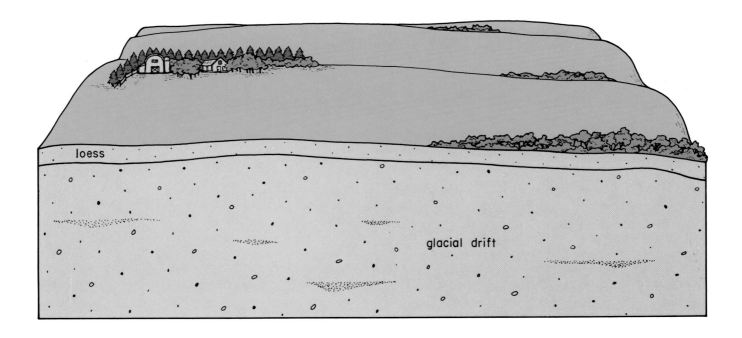

loess

glacial drift

continue eastward beneath the Des Moines Lobe, extending at least as far as western Franklin County where they appear in a quarry near Dows along the Iowa River. They also have been found just above water level in the deeply entrenched valley of Brushy Creek in eastern Webster County. The glacial advance(s) that deposited the Sheldon Creek drift took place just before Wisconsinan climates were at their coldest and just before massive loess deposition and intense erosional activity spread across Iowa's ice-free, tundra-vegetated landscapes.

Even though the eastern portion of the Northwest Iowa Plains contains these Wisconsinan-age glacial deposits, it and the rest of the region fell victim to the vigorous erosive activity that accompanied the glacial cold. In the process, most of the topographic signs of a freshly glaciated landscape were lost as streams lengthened and connected to form a well-established drainage network across the landscape. Today, the eastern portion of the

region no longer resembles the landscape of its glacial sibling to the east. Instead, the entire Northwest Iowa Plains region bears a strong similarity to the Iowan Surface in terms of erosional history and overall appearance. Keep in mind that the erosional events that leveled the Iowan Surface 16,500 to 21,000 years ago had an impact on all of northern Iowa. The effects of this erosional scouring in northwest Iowa remain apparent in the long, low, rolling swells and in the distant, uncluttered views to the horizon, even though the gentle slopes are mantled by loess (photo, above). As noted earlier, these effects in

north-central Iowa were later buried by the expressive glacial deposits of the Des Moines Lobe.

Glacial geology maps of the midcontinent usually show the Wisconsinan-age Des Moines Lobe advances as a composite outline of both the Sheldon Creek and the Dows formations (see map, p. 20). Mapping of the landform regions of Iowa, however, is based primarily on physical appearance or geomorphology, and this isolates the Des Moines Lobe region as only the more recent and more clearly observed of this two-stage Wisconsinan glacial activity in Iowa (see map, p. 31).

Open views across rolling landscapes to distant, uncluttered skylines are characteristic of northwestern Iowa. In this loess-mantled terrain, the highest and driest region of Iowa, wooded areas occur only along streams or as windbreaks planted around farmsteads.

Although there are similarities in the erosional histories of the Iowan Surface and the Northwest Iowa Plains, there are important differences in the age of the affected glacial deposits, as noted, as well as in the thickness of the loess mantle, the elevations of the land surface, and the present precipitation and vegetation. It is this combination of physical characteristics that unites the Northwest Iowa Plains and distinguishes this region from the state's other landform regions.

Windblown loess is abundant and nearly continuous across the region. The mantle of silt generally thins in a southwest to northeast direction from about 16 feet in Plymouth County to about 4 feet in Clay County. This depositional pattern reflects the region's proximity to the Loess Hills and to the Missouri and Big Sioux river valleys where this fine-grained material originated. The effect of the loess mantle in northwestern Iowa is to steepen some hillslopes and smooth out and fill in irregularities elsewhere in the landscape. In addition, the pattern of loess accumulation tends to obscure the region's boundaries with the Loess Hills to the southwest and the Southern Iowa Drift Plain to the south. The loess cover, however, does help to differentiate the Northwest Iowa Plains from the Des Moines Lobe which has no loess deposits on the land surface.

Added to these factors of geological history and materials are subtle differences that influence the appearance of this particular landform region, such as elevation. Northwestern Iowa is a definite topographic step upward to the High Plains of the Dakotas. The lowest land surface elevation in Iowa, 480 feet, occurs in the southeastern corner of the state where the Des Moines River empties into the Mississippi. From that point, elevations gradually climb to the north and west (see map, p. 111). Iowa's highest point of land, 1,670 feet, is in a farmer's feedlot on a knobby ridge of glacial drift at the edge of the Des Moines Lobe about four miles northeast of Sibley in Osceola County. While this high point is actually on the northwestern edge of the Des Moines Lobe region, altitudes throughout the Northwest Iowa Plains are uniformly higher than any other portion of the state.

Coinciding with these regional variations in elevation are similar changes in the distribution of woodlands and precipitation. Native tracts of timber decrease noticeably from eastern Iowa to western Iowa, and in the western part of the state, they continue to decrease from south to north. Except for trees planted as windbreaks around farmsteads or those confined to the more moist drainageways, the plains of northwest Iowa are barren of timber.

A corresponding relationship is found in the statewide distribution of precipitation. The highest mean annual precipitation occurs in southeast Iowa with 34 inches per year, and the rate of precipitation progressively decreases toward the northwest corner where the mean annual precipitation drops below 25 inches per year. Thus we are looking at a landform region which, in addition to its other attributes, is higher, drier, and less timbered than any other in the state.

Bedrock exposures, rare in this part of Iowa, are generally not a factor in the appearance of the landscape. Iowa's youngest sedimentary rocks, however, are widespread beneath the glacial deposits of the Northwest Iowa Plains. Cretaceous-age sandstone, chalk, limestone, and shale, usually concealed from view, are found in outcrops along the Big Sioux River valley in Plymouth and Sioux counties. Even younger Tertiary rocks are thought to underlie this region, as vertebrate fossils of definite Tertiary age, including rhinoceroses and three-toed horses, have been identified from Pleistocene sand-and-gravel deposits in western Iowa. The sharp edges and excellent preservation of these fossils indicate that they have not traveled far from their original burial site.

The various Cretaceous rock types include sands deposited about 100 million years ago in stream channels that drained southwesterly toward an advancing Cretaceous coastline. Marine shales and chalky limestones followed as expansion of the great interior seaway submerged the central United States from the western mountains to eastern Iowa about 90 million years ago. Vertebrate fossils from these strata in Iowa and nearby areas of South Dakota and Nebraska include skeletal fragments of fish, crocodiles, large marine reptiles (plesiosaurs, turtles), and sharks. The transitional marine, alluvial, and terrestrial environments also produced Cretaceous plant fossils, including cones and needles of co-

Iowa's oldest exposed bedrock is the reddish Sioux Quartzite. These durable quartz-rich rocks, over 1.6 billion years old, are protected at Gitchie Manitou State Preserve in northwest Lyon County. Some of the Precambrian outcrops are seen here at "Jasper Pool," an abandoned quarry dating from the 1800s.

Photo by Jean C. Prior

nifers as well as leaf impressions from trees resembling magnolia and sycamore. Microscopic spores and pollen of mosses and abundant ferns also have been found in the Cretaceous mudstones and lignites, carbon-rich deposits intermediate between peat and coal.

Bedrock also occurs at the land surface in the extreme northwest corner of Lyon County. Here visitors can glimpse exposures of the oldest bedrock that can be seen anywhere in Iowa—Precambrian-age Sioux Quartzite. This hard, durable formation has a uniformly pink to reddish color. The low ridges of colorful outcrops are protected as part of Gitchie Manitou State Preserve (photo, right). When examined closely, rounded grains of quartz sand tightly cemented together by silica tend to give the rock a glassy appearance. Stains of iron-oxide on surfaces of the original sand grains create the variety of pinkish to red colors. The layers of sand that compose this rock are sometimes arranged in sets of inclined or cross-bedded patterns. These reflect swiftly changing currents of shallow water that originally flowed in a complex of braided stream channels close to sea level between 1.6 and 1.7 billion years ago. The leading edges of many of

the smooth rock outcrops also display a high gloss from long exposure to the polishing action of dust-laden winds. Pale green lichens often grow over the more protected leeward rock surfaces.

Gitchie Manitou State Preserve contains the principal outcrops of Sioux Quartzite that are exposed along the banks of the Big Sioux River in Iowa. In 1870, geologist Charles A. White designated these exposures as the type locality of this formation. The ancient Iowa outcrops occur along the southern margin of the Sioux Ridge, a high rise of basement rocks that extends from west of Pierre, South Dakota, to New Ulm in southwestern Minnesota. Just north of Pipestone, Minnesota, a historically and archaeologically significant layer of red mudstone occurs within the beds of quartzite. This softer stone, also known as catlinite, had important spiritual significance to American Indian tribes of the Northern Great Plains who quarried and carved the red clay into beautiful ceremonial pipes and other important trade items. In 1839, Boston geologist Charles T. Jackson named the Indian pipestone catlinite in honor of George Catlin, the American artist who painted Indian life and first brought this area to the attention of whites after visiting the quarries in 1836. Some of the earliest European explorers and naturalists to set foot in northwestern Iowa were drawn up the Big Sioux River by reports of the quarries on the sacred Indian grounds along the river's banks.

In the very northwest corner of Iowa the state's youngest rocks (as presently mapped) rest directly upon the state's oldest rocks. Elsewhere these Cretaceous and Precambrian rocks are usually separated by thousands of feet of intervening Paleozoic deposits; here, however, the Paleozoic strata are very thin as they lap against the ancient Sioux Ridge. From central Sioux County on north, they are completely gone. Their absence probably results from a combination of some strata never being deposited and others being removed by erosion. The prominent, enduring Sioux Ridge must have been an imposing geographic feature of the Cretaceous landscapes and seascapes. It stirs the imagination to visualize the steep, rugged coastline that must have existed when the deepening Cretaceous seas crashed against the colorful headland cliffs of Sioux Quartzite.

One of Iowa's largest remaining glacial erratics is composed of this distinctive Sioux Quartzite. It lies near the southern border of the Northwest Iowa Plains, well southeast of its outcrop area. Pilot Rock, as this landmark is appropriately named, can be seen perched along the eastern margin of the Little Sioux River valley about three miles south of Cherokee. Smaller boulders of the reddish quartzite are profusely scattered in the Pre-Illinoian and Wisconsinan glacial drift of western Iowa.

The contact between the dense glacial tills and the more porous loess at the land surface in the Northwest Iowa Plains results in the occasional presence of springs and seeps. These features are sometimes seen along sloping land surfaces or valley margins that intersect groundwater movement along the contact with the less permeable till. As in the Southern Iowa Drift Plain, this hydrologic setting is occasionally tapped by shallow, large-diameter wells. Another potential source of well water is the water-saturated, sand-and-gravel layers within the deeper glacial tills. Many large-volume users, however, must rely on the more extensive and more dependable sand-and-gravel deposits that occur along the river valleys of the region. Where these alluvial sources are not available or experience the contamination problems to which these shallow sources are vulnerable, alternative groundwater supplies usually can be found in the deeper Cretaceous sandstones.

As with most of Iowa, the terrain and soils of this region are well suited to the cultivation of crops, with some pastureland on the steeper slopes. With the low moisture available in this part of the state, however, water-conservation measures are particularly important. The sand-and-gravel resources mentioned earlier are also valuable for highway construction and road maintenance. The durability of the Sioux Quartzite also makes it a good source of road aggregate and railroad ballast in this part of Iowa, as well as in neighboring Minnesota and South Dakota. This quartzite also has a long history of use as dimension stone (Sioux Falls "granite"), which is apparent in the number of buildings in Sioux City and elsewhere constructed of the decorative stone.

Subtle changes in the appearance of Iowa's landscape are observed as one approaches the northwestern part of the state. The reasons for these differences are not as easily identified as in some parts of the state. They result from a combination of factors which include an erosional history that left an integrated network of streams across the region, a nearly continuous mantle of loess blown from nearby valley sources of fine-grained glacial outwash, and climatic and elevational distinctions that produced a treeless prairie on landscapes that lead to the continent's High Plains.

Paleozoic Plateau

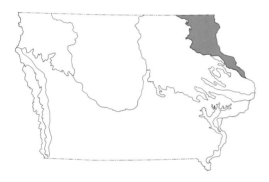

Then if he climbs to the nearest
commanding summit, he finds that
the maze of hills and the labyrinth
of ravines blend into a strongly
undulating plain, inclining gently,
though wrinkling deeply, toward
the Oneota [Upper Iowa]
and Mississippi . . .

—W. J. McGee
"The Pleistocene History of
Northeastern Iowa," 1891

If Iowa's landscape had to be divided into only two regions, one would be northeastern Iowa and the other would include everything else. The rugged, deeply carved terrain seen in the Paleozoic Plateau is so unlike the remainder of the state that the contrast is unmistakable, even to a casual observer. The most striking differences include abundant rock outcroppings, a near absence of glacial deposits, many deep, narrow valleys containing cool, fast-flowing streams, and more woodlands. This spectacular high-relief landscape (photo, p. 86) is the result of erosion through rock strata of Paleozoic age. The bedrock-dominated terrain shelters unusually diverse flora and fauna, including some species normally found in cooler, more northern climates. Samuel Calvin, one of Iowa's best known nineteenth-century geologists, spoke of these unexpectedly scenic landscapes as the "Switzerland of Iowa."

The key to this refreshing difference in appearance is the widespread occurrence of shallow Paleozoic-age sedimentary bedrock. In no other region of the state is bedrock in such complete control of the shape of the land surface. The Paleozoic strata include fossiliferous rocks of Cambrian, Ordovician, Silurian, and Devonian age which originated as sediments accumulating on sea floors and along coastal margins of tropical marine environments that existed here between 300 and 550 million years ago. In the course of geologic time, these deposits hardened into brittle rock strata which were later deformed and fractured by crustal movements within the Earth. A widespread series of vertical cracks now extend through these rocks; some of these weakened planes are

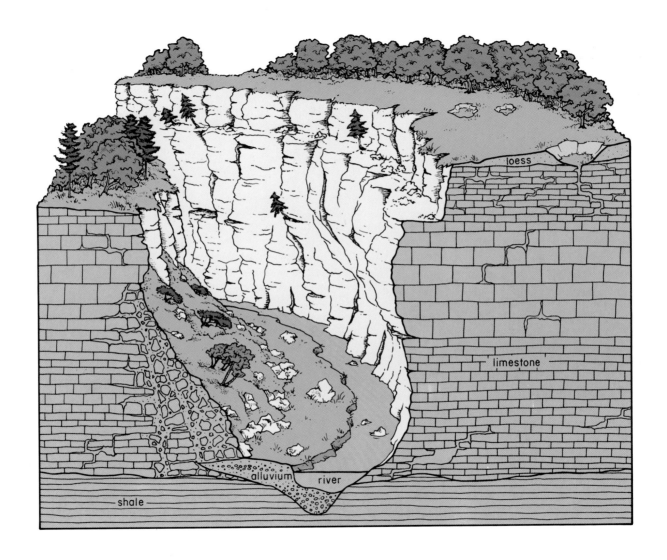

loess

limestone

alluvium   river

shale

85

Gary Hightshoe

parallel to each other, others are oriented at nearly right angles. These features, called joints, are responsible for the blocky shapes and sheer faces commonly seen along rock bluffs (photo, p. 89) and roadcuts. Their location also controls the abrupt turns noticed along many stream courses and valley segments throughout the region.

The entire sequence of Paleozoic rocks tilts gently toward the southwest, extending into deep basins in Kansas and Oklahoma. Ancient erosional processes in Iowa, including the work of Pre-Illinoian glaciers, beveled these inclined layers so that progressively older bedrock units appear at the land surface toward the northeast corner

The rugged landscapes of northeastern Iowa result from weathering and stream erosion of Paleozoic-age bedrock. These steep forested blufflands, underlain by resistant Ordovician dolomite, overlook the island-laced channel of the upper Mississippi River in Clayton County.

of the state. These sedimentary rocks are composed of limestone, dolomite, sandstone, and shale which vary in their resistance to erosion. Durable limestone, dolomite, and sandstone formations stand out as cliffs, pinnacles, ledges, and bluffs high on the landscape or as waterfalls and rapids breaking the stream flow along valley floors. Shale, a less resistant rock, usually produces smoother, more regular landscape slopes. The different erosional characteristics of various strata result in terrain that mirrors the composition of the bedrock sequence so closely it is often possible to trace distinct geologic formations across the landscape. Broad, sweeping views of uniformly rolling summits often appear connected to lower but similar levels or to valley floodplains by steep escarpments. It is these different landscape levels that produce the angular stepped skyline and the distinctive plateaus in this part of Iowa.

The most easily observed Paleozoic strata protrude along the prominent winding line of wooded bluffs marking the Silurian Escarpment. This high, steep, nearly continuous rim of east-to-northeast-facing rock follows an irregular line from central Fayette County to southeastern Jackson County, where the sudden drop in the landscape also defines the western and southern borders of the Paleozoic Plateau. The plunge from the gently rolling Iowan Surface into the deeply carved, high-relief landscapes of the Paleozoic Plateau is clearly seen along the northwest-to-southeast line connecting West Union, Fayette, Strawberry Point, Edgewood, Colesburg, and Peosta. This bluffline is interrupted by numerous small, narrow gorges and steep ravines which cause abrupt local changes in the direction of exposure. These various slope positions provide abundant havens of cool, moist, wooded habitats rich in diverse and sometimes rare communities of plants and animals. A noticeable concentration of protected natural areas in Iowa occurs along this prominent physiographic feature, including Brush Creek Canyon, Mossy Glen, Bixby, and White Pine Hollow state preserves.

The Silurian Escarpment is actually the leading up-dip edge (cuesta) of the massive slabs of Silurian-age dolomite and limestone that slope gently back to the southwest across Iowa. These strata form the uppermost bedrock across an area at least the width of a county before they are overtaken on the west by younger layers of Devonian limestones (see map, p. 17). Before the Mississippi Valley came into existence, Silurian strata also continued to the east. Today, however, only flattopped remnants of these rocks, termed outliers, are seen capping steep, isolated hills. Sherrill Mound in Dubuque County is a good example of an erosional outlier that is still close to the escarpment. Sinsinewa Mound located in southwestern Wisconsin and Scales Mound in northwestern Illinois are other Silurian outliers visible on a clear day from high elevations in northeastern Iowa, especially

87

Balltown Ridge in Clayton County. The Silurian Escarpment returns again in northwestern Illinois, then snakes northward to rim the western and northern shores of Lake Michigan, the northern shore of Lake Huron, and the southern shore of Lake Ontario, reemerging spectacularly as the resistant limestone ledge that forms Niagara Falls at the eastern end of Lake Erie. The prominent bulge in Iowa's eastern border begins where Silurian rocks cross into Illinois at Dubuque County. This eastward deflection of the Mississippi River further demonstrates the resistance of these rocks to erosion.

Immediately below the Silurian rocks defending the escarpment in Iowa are more gentle slopes developed on older, soft, greenish gray Ordovician shale (Maquoketa Formation). This change marks a significant geologic boundary between Silurian and Ordovician time periods and between contrasting rock types of dolomite and shale. The zone of contact is often observed as a sharp break in topography from nearly vertical slopes to more gently inclined hillsides. The change in slope gradient also is often recognized by a noticeable change in land-use patterns from timbered Silurian slopes to pasture or row crops on the Ordovician. Seeps and springs are common features along valley sides that intersect the geologic contact between the highly permeable dolomite and the less permeable shale beneath. Additionally, angular slump blocks of dolomite, pitted and gray from long exposure to the weather, are often seen in tilted positions on the lower pastured slopes of shale where they have slid from their original outcrops. Balltown Ridge and county road X-71 northeast of Luxemburg in Dubuque County,

Chicken Ridge in Clayton County, and Goeken Park (capped with Devonian rocks) overlooking the town of Eldorado in Fayette County are excellent places to observe some of these bedrock relationships and the very scenic landscapes they produce.

Continuing northeast from the Silurian Escarpment and underlying Ordovician shales are other plateau-like surfaces upheld by still older Ordovician limestone and dolomite strata (the durable Galena and Prairie du Chien groups). These rolling bedrock-controlled uplands also display massive escarpments that descend into or toward the great trench of the Mississippi Valley. The precipice where the Galena Group dolomites drop directly into the Mississippi Valley is especially picturesque at the Mines of Spain south of Dubuque, at the scenic overlook along U.S. 52 south of Guttenberg, and at Pikes Peak State Park south of McGregor. Mount Hosmer at Lansing and Blackhawk Bluff south of New Albin are other escarpments formed by still older Cambrian sandstone (Jordan Formation). Another scenic area where we can observe the effects of different bedrock formations on the landscape is found along the ridge-crest roads that radiate north and east from the Allamakee County highlands in the vicinity of Waukon. These vantage points overlook rolling landscapes with descending escarpments or steps formed by successively older outcrops of Ordovician dolomites (Galena), sandstone (St. Peter), older dolomites (Prairie du Chien), and eventually Cambrian sandstone (Jordan).

The limestones and dolomites that dominate so many of the Paleozoic Plateau landscapes are carbonate

Jean C. Prior

Horizontally layered sedimentary strata, in combination with fractures and crevices that extend vertically through the rock, contribute to the angular, blocky appearance of this weathered dolomite outcrop along a gorge-like tributary of the Turkey River in central Clayton County.

rocks. These lime-rich strata are slowly soluble in groundwater percolating through them from the land surface. The result of this dissolving action is the gradual enlargement of cracks, crevices, and other zones of weakness into caves and cavern systems. When these underground passages extend too close to the land surface, thin bridges of soil and rock collapse to form sinkholes. Concentrations of sinkholes can be seen in many parts of the region, including the vicinity of Iowa 9 in western Allamakee County and the area east of U.S. 52 in southern Clayton County (photo, left). Even streams are occasionally swallowed by sinkholes in channel floors; this disappearance of flow can leave blunt-ended or abandoned valleys on the land surface.

Springs are another sign of subterranean drainage systems. Abundant in this region of Iowa, they usually are found along steep valley sides where erosional deepening has intercepted the flow of groundwater. Even during long periods without rain, streams in this region continue to flow because of the significant contribution of groundwater from bedrock aquifers. Dunnings Spring and Twin

The presence of shallow limestone beneath the landscape is revealed by this dimpled terrain in Clayton County. The circular depressions, some filled with water or clusters of trees, mark the collapse of soil and rock material into subterranean crevices and caves. These sinkholes develop from the dissolving action of groundwater.

Photo by Gary Hightshoe

Springs, municipal parks in Decorah, and Cold Water Spring State Preserve in Winneshiek County are picturesque examples of these intriguing geologic and hydrologic phenomena. Cold Water Spring is the outlet of underground flow from one of Iowa's largest cavern systems, Cold Water Cave (photo, p. 92). The best place to see and experience caves in Iowa, however, is in Jackson County at Maquoketa Caves State Park, along the northeastern edge of the Southern Iowa Drift Plain.

Winter air, pulled by gravity into northeastern Iowa's sinkholes and creviced limestones, can cool the subterranean fissures well below freezing. When spring snowmelt and rainfall move down into crevice systems, ice sometimes builds up on the rock walls in sufficient thicknesses to remain well into summer. These remarkable ice caves then continue to refrigerate the air circulating through them during warmer seasons. When intercepted by steep slopes, the cold-air discharges behave like springs, bathing lower slopes in a continuous flow of cool, moist air. Decorah Ice Cave, a well-known state preserve along the Upper Iowa River valley in Winneshiek County, is a good example of these processes. These unusual microclimates, combined with steep north-facing slopes and abundant rock talus at their bases, can produce unique cold-air (algific) slopes, a particularly rare and sensitive biological habitat in Iowa.

The appearance of sinkholes, springs, and caves in landscapes underlain by shallow carbonate bedrock is known as karst topography. This geologic condition is the reason behind many of the region's interesting landforms and unusual biological habitats, as well as its high vulner-

Caves are a fascinating aspect of karst landscapes. This time-lapse photo taken inside Winneshiek County's Cold Water Cave shows actively growing whitish formations of calcium carbonate, including glistening mounds of flowstone (top right) and intricate "draperies" (top left). The lime-charged groundwater seeps in through cracks crisscrossing the cavern ceiling.

ability to groundwater contamination. Karst features in Iowa are not restricted to the Paleozoic Plateau, but they do occur here most frequently and are a distinguishing characteristic.

In addition to the strong visual impact of various types of bedrock and karst features on the landscape, one other obvious factor distinguishes the Paleozoic Plateau landscapes—the character of its river valleys. The region's largest valley, the gorge of the Mississippi, is one of the most prominent physiographic features in the central United States. Tributaries that spill eastward into the Mississippi Valley over steep gradients through the Paleozoic Plateau include the Upper Iowa, Yellow, Turkey, and Volga rivers. The river valleys are deep, narrow, steep-sided corridors carved into bedrock, often in surprisingly tight meander patterns. Such entrenched valleys usually display strong bedrock control of their courses, including abrupt, sharp-angled turns that trace the embedded fracture patterns described earlier.

As these river valleys evolved, some of their meander loops were abandoned, leaving horseshoe-shaped segments of old valley floors perched above existing floodplains. The abandoned meanders often partially surround steep, isolated rock cores. Sand Cove, standing more than 100 feet above the present floodplain along the south side of the Upper Iowa Valley in Allamakee County, is an excellent example. "The Elephant" and Mount Hope, nearby along the north side of the Upper Iowa Valley, are examples of large rock cores positioned on the inside bends of abandoned valley segments.

Another impressive abandoned valley slices through

the landscape just north of Dubuque (photo, pp. 94–95). This abandoned channel, known as Couler Valley, once carried the ancestral flow of the Little Maquoketa River to its junction with the Mississippi Valley near the present site of downtown Dubuque. The thin, steep divide near Sageville that separated their valleys was breached, and the Little Maquoketa waters were captured and permanently diverted along the shorter, more direct route straight east through Peru Bottoms. Horseshoe Bluff at the Mines of Spain also separates an ancestral valley of Granger Creek and Catfish Creek from the Mississippi. These scenic, abandoned valley segments and abandoned rock-cored meanders are a fascinating aspect of valley evolution in the Paleozoic Plateau.

Another component of northeastern Iowa valleys are the thick alluvial deposits within them. In the case of the upper Mississippi, large meltwater floods during the waning phases of the Wisconsinan scoured the valley, removing or rearranging earlier records. Side valleys now store the geologic ledgers registering earlier fluctuations in river level and sediment load that occurred out in the master valley. The ledgers exist in the form of alluvial terraces, those level but elevated remnants of older floodplain surfaces that flank valley margins. Clear Creek and Village Creek valleys west and southwest of Lansing contain well-preserved terrace remnants as high as 60 feet above the present floodplains. Terraces are important in documenting the region's complex alluvial history and the response of tributaries to changing water levels associated with glacial melting and drainage diversions in the Mississippi Valley.

For many years the eastern half of the Paleozoic Plateau was thought to be untouched by Pleistocene glacial activity and was referred to as the "Driftless Area," which included similarly rugged landscapes in the adjacent parts of Illinois, Wisconsin, and Minnesota. This concept is no longer valid, however, as patchy remains of Pre-Illinoian glacial drift over 500,000 years old are documented on stream divides in the Iowa portion of this area. Pre-Illinoian drift is even more widespread in the western half of the Paleozoic Plateau, although it still plays no significant role in the present appearance of the region's landscapes. Regardless of the relative amounts of glacial drift, the region stands united by the strong influence of bedrock geology, by a marked contrast to the remainder of the state where landscapes are subdued by glacial deposits, and by its many topographic and ecologic similarities.

Our understanding of Iowa's glacial-age landscape evolution has grown significantly in recent years. As specific geographic areas and stratigraphic deposits have been examined in detail, we have seen an emerging unity among the state's diverse landform regions. The Paleozoic Plateau, for all of its spectacular differences, fits neatly into the erosional history we have been compiling since looking at the Southern Iowa Drift Plain. Though the sedimentary rocks of this northeast Iowa landscape are ancient in comparison to all other landscape-forming materials in Iowa, the "preglacial" antiquity previously assigned to the landscape features themselves is being challenged. Recent investigations suggest that much of this rugged terrain is actually quite young. Stream erosion and hillslope development since Pre-Illinoian glacial

Drainage patterns of rivers are emphasized in this color-infrared aerial photograph of the Dubuque area. Landmarks include the Mississippi River, Lock and Dam 11, and the Little Maquoketa River with its ancestral channel, Couler Valley (seen in a straight southeasterly route through the city). The topography strongly influences land-use patterns. The false-color technique enhances differences between forest (dark green), pasture/alfalfa (pink), crop stubble (tan), and plowed ground (gray-green) in this November 1980 view from 40,000 feet.

Iowa Department of Natural Resources
Geological Survey Bureau

events have produced the deeply dissected landscape and in the process have removed the glacial deposits from all but isolated upland positions. The same episodes of erosion reflected in the changing hillslope gradients of the Southern Iowa Drift Plain and in the scoured Iowan Surface and Northwest Iowa Plains also had a significant impact on the Paleozoic Plateau. The effects, however, are magnified by the bedrock-controlled relief and complicated by the steep slopes. The isolated distribution of Pre-Illinoian glacial deposits found on stream divides high in the landscape shows that the Mississippi River evolved during the early to mid-Pleistocene. The river's network of tributaries in Iowa became established after the last Pre-Illinoian glacial events in the region 500,000 years ago. Loess-mantled, rust-colored Late-Sangamon paleosols are found on these remnant glacial deposits, and in places they extend onto the bedrock surface in the form of waxy red clays. At other locations loess lies directly on bedrock without any hint of a paleosol, conditions typical of the Wisconsinan-age Iowan Surface. These relationships confirm that multiple erosional episodes have cut across the entire Paleozoic Plateau.

Karst conditions develop at or just above the zone of groundwater saturation. Caves and passageways thus are strongly linked in time to periods when stream downcutting lowers the water table in shallow carbonate bedrock. The speleothems (stalactites, stalagmites, and flowstone) that decorate cavern systems cannot form until the karst openings fill with air and are largely free of groundwater. Uranium-thorium series radiometric dating of speleothems in Iowa and Minnesota indicate that major valleys were carved deeply enough into the landscape to bring the water table down and foster a major period of speleothem growth between about 100,000 and 160,000 years ago.

Most of the deep entrenchment of the region's valleys, as well as formation of the prominent rock-cored meanders, occurred during the Wisconsinan, before 30,000 years ago. The ancestral valley meanders filled with stream deposits and were abandoned between 30,000 and 20,000 years ago. Intense frost action, producing unstable slopes and massive mechanical weathering and erosion, was brought on by the severe cold of glacial climates about 21,000 to 16,500 years ago. Locally, these periglacial conditions produced massive colluvial slopes of angular, blocky rock fragments (talus) along the sides of entrenched valleys. Karst conduits near the land surface also collapsed into large, blocky talus. The small, narrow gorges, described earlier emerging from the Silurian Escarpment, often mark collapsed sections of former cavern systems that were exhumed by the intense late-Wisconsinan erosional episode.

From about 18,000 to 9,500 years ago, the waning Wisconsinan ice front fluctuated on either side of the continental divide in Minnesota, separating drainage to the Mississippi from that to Hudson Bay. Catastrophic meltwater floods periodically surged down through the Mississippi Valley. Left behind are the series of elevated terrace deposits mentioned earlier, composed of silts and clays deposited in the slack backwaters of the Mississippi's tributaries. For the most part, slower-paced sedimentation, minor downcutting, and lateral migration of river channels have been the dominant activities in the region's valleys since then. It is important to realize that the progression of deposits from these floodplains and terraces up to the steep colluvial slopes, to the filled and abandoned meanders, to the ornamentation in caves, and to the rusty red, clay-rich upland paleosols are important linkages in time that verify the surprising geologic youthfulness of these rugged landscapes, even though they are developed on ancient Paleozoic bedrock. The deposits also have important geographic links, demonstrating that many of this region's significant features were affected by the deepening presence of the Mississippi Valley.

Quarries are common in the limestone and dolomite deposits throughout the Paleozoic Plateau. The easily accessible bedrock is used for road construction and maintenance, though some is quarried for building stone. Bedrock deposits of the region also have a colorful mining history. Lead (galena) ore was mined for at least 300 years from veins occurring along crevices in the Ordovician (Galena Group) dolomites exposed along the Mississippi Valley and its tributaries. The hills around Dubuque are honeycombed with shafts and adits of this lead-mining activity. Even iron ore was mined in the 1890s from a small, unusual occurrence of Cretaceous rocks found northeast of Waukon. This area, known as Iron Hill, is one of the highest points of land (1,345 feet) along this segment of the upper Mississippi Valley.

In addition to the economic value of the bedrock materials, many water wells have been drilled into the shallow creviced limestones and porous sandstones for drinking water. The karst characteristics of the area are a vivid reminder that contaminants from the land surface have direct access to these groundwater zones. In this vulnerable geologic setting, care must be taken to manage waste disposal and land-applied farm chemicals to prevent pollution of underground aquifers. Widespread contamination of the region's shallow aquifers by nitrate and bacteria already has caused many wells to be abandoned and new wells to be drilled to deeper groundwater sources.

The Paleozoic Plateau also contains a large share of the state's native woodlands. The steep, rocky slopes are unsuited to cultivation and remain forested. Some of this timber cover actually has developed since the time of Euro-American settlement when naturally occurring fires were suppressed. Along steep, dry slopes above major valleys, remnant prairies called goat prairies often can be spotted among the forested slopes.

With cropland at a premium in many areas, the small but productive floodplains and terraces along valleys are usually farmed. Areas of level uplands are also cultivated or used as pasture, especially for dairy cattle. The rugged topography lends itself to many types of outdoor recreation in both winter and summer. Those fishing for trout seek the deep, quiet pools below rocky riffles along the numerous spring-fed streams. Canoe enthusiasts can travel many miles of wooded and rock-bound waterways. Collectors can find abundant fossils in many of the exposed bedrock formations. The heavily timbered landscape supports a colorful autumn foliage, and the scenic region attracts many visitors each year.

Alluvial Plains

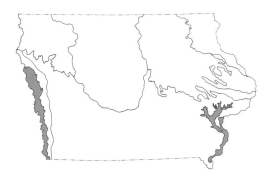

. . . the broad "Father of Waters" meandering through a densely wooded flood plain . . . in channels and sloughs, together cutting the river bottom into a labyrinth of land and water, island and lake, sandbar and marsh, tangled brake and boiling chute, which he could scarce traverse either afoot or afloat without a guide.

—W. J. McGee
"The Pleistocene History of Northeastern Iowa," 1891

Water in motion is one of those timeless processes in nature that hold our gaze and set our thoughts free. For landlocked Iowans, far removed from the rhythmic sight and sound of ocean surf, the strong silent flow of a large river has its own soothing and enduring attraction. Iowa also has hundreds of small brooks and creeks that can be crossed with a good long-jump or by convenient stepping-stones. It is time well spent to look closely at these large and small drainageways, for they demonstrate the most significant geological process presently at work on the Iowa landscape—the action of water. Rivers construct distinctive, flat-floored corridors known as alluvial plains which are underlain by water-transported deposits. These topographic corridors weave throughout the state's other landform regions, but together they constitute the last of Iowa's seven physiographic regions, the Alluvial Plains.

During the tens to hundreds of thousands of years since the various Pleistocene glaciers melted from Iowa, rivers have carved the state's valleys and partially filled them with layered deposits of gravel, sand, silt, and clay. The streams draining Iowa's land surface today range in size from small rills on upland slopes to the broad Missouri and Mississippi lowlands along the state's western and eastern borders. Only these largest segments of the Alluvial Plains region stand out at the scale of the map on page 31. These drainage networks transfer water and sediment from the highest portions of the watershed downslope through increasingly larger streams and rivers and eventually to the Mississippi delta in the Gulf of Mexico.

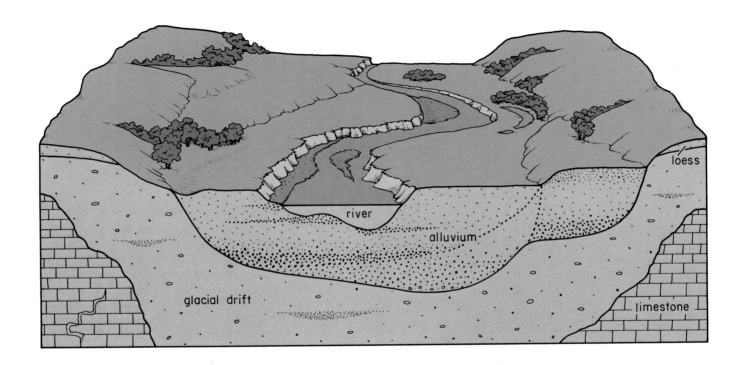

Along shallow stream channels, it is easy to observe individual particles of sediment being swept quickly downstream through narrow chutes. Sediment grains may lodge temporarily in a quiet pool or at the downstream end of a rippled shoal of other grains. It is also common to see the outer bends of meandering stream channels undercut their steep banks, causing material to slump into the flow, while sediment accumulates along the shallower inside curves of bends, forming point-bar deposits (photo, p. 100). Neither sediment nor water moves at a uniform pace through river systems. More water and stronger currents can carry greater amounts of larger-diameter materials such as pebbles, cobbles, and boulders over longer distances. Lower velocities cause deposition of the coarser load and transport of only finer-grained particles of silt and clay. Within river systems,

Flowing water shapes valley sides and floors while continually eroding and depositing loose sediment. Here the Turkey River, meandering across its floodplain in Clayton County, undercuts steep banks on the outside of bends, while crescent-shaped point-bar deposits accumulate on the shallower inside curves.

episodes of downcutting and sediment transport alternate with lengthy periods of sediment storage on both small and large scales and over short as well as long periods.

The network of pathways that rivers have carved into Iowa's land surface is seen on the topographic relief map (see p. 33). Landscape features resulting from the action of rivers appear along the Missouri and Mississippi valleys as well as along all of the state's interior streams. Each valley contains a collection of landforms that record portions of the river's geologic past, including evidence of large-scale horizontal and vertical changes within its valley.

Floodplains, as the name implies, are nearly level

though often uneven lowlands adjacent to a river channel. The plains are submerged when a river channel carries excess water, as often occurs in the spring after snowmelt or heavy rains. In river valleys, active erosion and deposition take place on floodplains, and valley floors are often scarred with low ridges and swales marking former positions of a river channel (see photo, p. 8). As a river meanders across its floodplain, channel cutoffs occur, sometimes leaving isolated crescent-shaped bodies of water known as oxbow lakes in the abandoned channel loops. Backwater sloughs and stands of timber frequently indicate these poorly drained meander scars. These various landform features, while present along many Iowa rivers, exist on a grand scale within the Missouri and Mississippi valleys. The curved intermingling of land and water along the lengths of these broad alluvial corridors explains their important role as flyways, hosting hundreds of thousands of migratory waterfowl each spring and fall. The watery lowlands also support other riverbank dwellers such as beaver and muskrat as well as a wide variety of wetland plant communities. DeSoto Bend National Wildlife Refuge in Harrison County and Lake Odessa in Louisa County, part of the Mark Twain National Wildlife Refuge, are good examples of ecological habitats supported by alluvial plains.

The geologic restlessness of rivers is also preserved in historical documents. When maps of the Missouri River drawn by the Lewis and Clark expedition in 1804 are compared with later steamboat company and government surveys, they show an animated river, its floodplain etched with intricate scrolls produced by shifting loops of the river's course. Considerable political and legal strife has occurred because state boundaries and landownership records were based on the position of the Missouri River at a single date in history.

Nor do river channels and floodplains remain vertically stationary within their valleys. Evidence for these large-scale changes is recorded in terraces and benches along valley sides and in the depth of alluvial deposits that backfill a valley's original trench, buried beneath the present floodplain (photo, p. 103). Terraces and benches are nearly level surfaces but are elevated above existing floodplains, usually by a distinct slope or scarp. They are remnants of earlier floodplains, abandoned when a river began a new episode of downcutting. Like floodplains, terraces are composed of materials transported by a river—stratified assortments of clay, silt, sand, and gravel. These generally porous deposits allow easy infiltration and subsurface movement of water. Benches, on the other hand, while looking like terraces and veneered with alluvial deposits, are underlain by a shelf of older material which is not alluvium but is usually bedrock or glacial drift that resisted the river's downcutting. Multiple terraces and benches in a valley reflect the complexity of a river's history, as each lower topographic level represents a more recent step in time as the river eroded its valley more deeply. Fine examples of these features occur along the Des Moines Valley between Fort Dodge and Des Moines, especially the Brushy Creek valley east of Lehigh in southeast Webster County.

Small tributaries entering a main valley are often steep-gradient, intermittent streams. During episodes of discharge from uplands out onto the more level flood-plain, terrace, or bench, these small streams lose their capacity to carry sediment. Fan-shaped aprons of alluvium accumulate at the abrupt change in slope, burying older floodplain materials and soils beneath them. Alluvial fans frequently can be identified along valley margins by the interruptions in field patterns reflecting subtle changes in slope and drainage.

While most of the landforms and deposits found on alluvial plains reflect the effects of flowing water, the influence of wind is also seen. During seasonal low-flow conditions along stream channels, alluvial deposits emerge as sandbars. If not anchored by vegetation, sand and silt can be blown onto floodplain and terrace surfaces as well as onto higher elevations along valley margins. These accumulations of dune sand appear along terrace edges or fringe the uplands of major valleys where they are often held in place by stands of oaks. Sand-dune topography occurs downwind of broad reaches of valley floor, as along the Mississippi and segments of the Des Moines, Skunk, Iowa, and Cedar rivers in eastern Iowa. Rochester Cemetery in Cedar County, known for its colorful spring wildflowers and stately oaks, and Marietta Sand Prairie, a state preserve in Marshall County, both host native plant species in upland sand-dune habitats. In occasional hollows or depressions (blowouts) among the dunes, ponds form where drainage is sealed by underlying clays; these wetlands are another unusual and limited habitat in Iowa. Behrens Ponds and Woodland State Preserve in Linn County is a good example; Rhexia Pond in Muscatine County is another. A few dune fields

are active today, shifting seasonally with the winds, as at Sand Cove on a high terrace of the Upper Iowa River in Allamakee County and at Big Sand Mound, a terrace remnant in the Mississippi Valley straddling the Muscatine-Louisa county line. Most dune tracts in Iowa are presently stabilized; they originated during late Wisconsinan time, and many were reactivated in response to the warmer, drier climates existing 4,000 to 8,000 years ago during an arid portion of the Holocene called the Hypsithermal, the period of maximum postglacial warming.

The size, shape, and complexity of alluvial plains landforms depend on the individual river's past experience, including the age of its valley, the long-term variations in available water and sediment, the geologic material into which the valley is carved, and fluctuations in the level of the water body into which the river flows. These factors control valley characteristics and affect landforms along alluvial plains throughout the state.

The youngest valleys in Iowa are found eroded into the low-relief landscapes of the Des Moines Lobe. Compared to others in the state, these valleys have barely begun to excavate a place for themselves in the landscape. As the Des Moines Lobe glacier melted from this area between 14,000 and 11,000 years ago, only a few large valleys developed—the Boone, Raccoon, Ocheyedan, Winnebago, and the upper reaches of the Des Moines, Skunk, Iowa, and Little Sioux. These valleys were excavated by large volumes of meltwater, and their older terraces contain high-energy accumulations of coarse sand and gravel originating from melting ice and left in place by valley downcutting. Many of Iowa's rivers appear too

Rivers excavate valley corridors through uplands of older and different geologic materials. The Upper Iowa River has carved a winding course through the bedrock uplands of the Paleozoic Plateau in western Allamakee County. Its valley contains terrace remnants of earlier floodplain levels (rimmed by trees along inside of bend).

small to have excavated the valleys in which they flow. These underfit streams and oversized valleys are clear evidence of the enormous volumes of glacial meltwater that once flowed through them.

A well-integrated network of drainageways extends across the more rolling and steeply sloping landscapes of the remaining regions in northwestern, southern, and northeastern Iowa. Valleys in these regions generally are carved into older glacial materials deposited more than 500,000 years ago, and their older terraces often are mantled with late-glacial deposits of loess, indicating an age greater than 12,500 years. Western Iowa streams are particularly noted for their deep gullies carved into thick, easily eroded loess deposits. Valleys in extreme northeastern Iowa are also distinct, with narrow gorges entrenched deeply into bedrock. These valleys, however, were cut in response to the deepening Mississippi Valley, not by glacial meltwater draining through them.

The Missouri and Mississippi alluvial plains are distinct among Iowa's valleys because of their great size and the fact that they owe their origins to events which occurred outside the state's boundaries (photo, left). Both valleys carried large volumes of glacial meltwater in their headwaters between 30,000 and 9,500 years ago; in addition, they were principal sources of the loess mantling much of the Midwest. The Mississippi Valley's geographic alignment today is a compromise, determined in part by ancestral ice-marginal streams accompanying Pre-Illinoian glacial advances out of the northwest over 500,000 years ago and the Illinoian advance from the northeast about 150,000 years ago. The valley along Iowa's border was finally knit together by massive meltwater drainage diversions that excavated a new channel from Clinton to Muscatine (the Port Byron Gorge) between 25,000 and 21,000 years ago. The geological evolution of the Missouri Valley is less well known, but evidence shows it shared a similar history. Rivers with headwaters in the central Rocky Mountains once flowed eastward across western Iowa, perhaps all the way to the ancestral Mississippi system. These waters were diverted southward by Pre-Illinoian glacial advances into the region, and the Missouri Valley's final position probably was established by or during Illinoian time.

The distribution of major rivers in Iowa, as well as the drainage divide that separates the watersheds of the Missouri and Mississippi rivers, is shown on page 32. The Mississippi-Missouri basin is one of the world's largest drainage systems. Rivers in the western one-third of Iowa drain southwestward to the Missouri while those in the remainder of the state drain southeastward to the Mississippi. This major divergence of Iowa's surface-water flow toward opposite ends of the state suggests the presence of some prominent topographic ridge; however, the divide is seldom sufficiently elevated above the surround-

The broad, level, cultivated alluvial plain of the Missouri River ends abruptly against the steep irregular loess bluffs in western Iowa. Thick sand-and-gravel deposits beneath the floodplain are an important source of groundwater supplies for wells.

Photo by Mark Engler

ing land surface to be distinguished by the eye. Three-fourths of a mile east of Arcadia, along U.S. 30 in Carroll County, is a marker that notes the crossing of this boundary, one of the Midwest's most significant yet modest geographic thresholds.

North of Cherokee County is an obvious deviation from the divide's northwesterly trend. The waters along this portion of the Little Sioux basin, including the Ocheyedan River, once flowed southeast to the Mississippi system. This drainage was diverted about 14,000 years ago when the Des Moines Lobe glacier blocked the rivers, forming a large but short-lived ice-marginal lake at Spencer in Clay County. Glacial Lake Spencer eventually overflowed; its waters sought a route around the ice blockage, scouring a new canyon-like valley between Sioux Rapids and Peterson in Buena Vista County (see map, p. 6) and connecting westward with the ancestral Little Sioux River, a tributary of the Missouri. This drainage became permanent, doubling the size of the Little Sioux River basin. Its former orientation, however, is still noticeable in the altered alignment of the state's major drainage divide.

The matter of ice-marginal lakes is also tied to past interpretations of the Lake Calvin area in Louisa and Muscatine counties. This prominent lowland between the converging Iowa and Cedar rivers was regarded in the early geological literature as the site of a large lake formed when the two rivers were blocked by the westward advance of Illinoian ice into southeast Iowa. Such drainage disruptions undoubtedly occurred; prominent diversion channels of the Mississippi can be seen today as broad swales crossing the eastern Iowa uplands in Clinton County (Goose Lake Channel) and Scott and Muscatine counties (Cleona Channel). Unresolved, however, is the location for a diversion channel south of the Iowa-Cedar drainage. Researchers now suggest the combined drainage may have flowed south beneath the glacial ice. At any rate, stratigraphic studies and radiocarbon dating indicate the landscapes within the Lake Calvin basin, formerly thought to be remnants of Illinoian lake beds, are actually composed of much younger Wisconsinan and Holocene deposits. The floodplains, terraces, dunes, and thick alluvial deposits which characterize this area are the result of complex alluvial origins. If continuous lake deposits existed, they have since been buried, reworked, or removed by erosion.

Alluvial processes, easily observed but seldom appreciated for their geologic significance, have been at work for much of the Earth's history. For example, ancient river channels now filled with sandstone are part of Iowa's sedimentary rock records, especially those of Pennsylvanian and Cretaceous age. They are evidence of river courses through ancient landscapes, flowing to oceans of which only fossils and rock remain. Good exposures of these alluvial channel sandstones can be seen at Wildcat Den, Ledges, Dolliver, and Springbrook state parks. Younger valleys carved into Iowa's bedrock were buried by various Pleistocene glacial deposits and are now hidden from view. Some of these buried valleys contain deposits of alluvial sand and gravel sought by well drillers for groundwater supplies.

The deposits of sand and gravel associated with alluvial plains statewide are important sources of groundwater. The capacity of alluvial materials to yield water to wells varies along a valley's length. These variations are caused by changes in the deposits, which reflect their past environments of deposition. The source of groundwater in these alluvial aquifers is precipitation and runoff which percolate through soils. Groundwater levels in alluvial deposits change throughout the year, depending on the frequency and intensity of rainfall. A river level usually indicates the position of the local water table or the top of the zone of groundwater saturation, as do many of the oxbow lakes and sloughs on an adjoining floodplain.

The network of alluvial plains established through the Iowa landscape by rivers and their valleys has had a significant impact on patterns of human activity. These natural corridors offered pathways along which travel and settlement took place in both historic and prehistoric times. Archaeological sites are concentrated along bluffs and terraces of large valleys. Effigy Mounds National Monument as well as Fish Farm Mounds, Turkey River Mounds, Malchow Mounds, Little Maquoketa Mounds, and Toolesboro Mounds state preserves are all prehistoric earthworks that overlook or are within the Mississippi Valley.

Episodes of alluvial fan development along the state's valleys overlapped periods of prehistoric occupation, resulting in the geologic burial of many archaeological sites. Landscape evolution studies of Iowa's alluvial plains thus are providing important new information to archaeologists and natural-resource planners about possible preservation of other significant but less obvious prehistoric cultural remains in Iowa.

Marquette and Jolliet, the first explorers of European descent to make contact with the American West, arrived in the upper Mississippi Valley via the Wisconsin River in 1673 at what is now Pikes Peak State Park in Clayton County. By the time of the American Revolution in 1776, the lead and zinc deposits mined along the Mississippi Valley were an important national resource. Julien Dubuque's tract, known as the Mines of Spain, marked the beginning of permanent Euro-American settlement in Iowa. Nearly all of Iowa's principal cities and towns grew from settlements along river valleys. The routes of early exploration became indispensable arteries of supply, travel, and commerce as the state developed.

Though alluvial plains occupy a small percentage of Iowa's land area, increasing demands are being placed on them. In addition to urban and industrial use, valleys support intensive agriculture, expanding recreational activities, increased pumping of groundwater for urban, industrial, and irrigation uses, extraction of sand-and-gravel resources, and important wetland habitats—and they continue to perform their geologically designed function of accommodating floodwaters. The hazards to people and property from these recurring geologic events should be a basic consideration of all land-use decisions in valley landscapes. In addition, this intense utilization, combined with a shallow, unprotected groundwater resource, makes alluvial plains especially vulnerable to contamination.

Visits
to Iowa Landforms

Clambering down through a tangle of ferns and wild flowers one reaches the bed of a small stream just below where it tumbles twenty feet over a lichen-covered ledge. The steep walls, dotted with mosses, harebells and rock ferns, rise to such height as to exclude all save the noonday sun and bury the gorge in fragrant coolness.

—A. G. Leonard
"Geology of Clayton County," 1906

It is a valuable and exciting experience to travel to new places or revisit old favorites accompanied by knowledge about the landscape and its origins. Vacations, holidays, and field trips lure people to out-of-the-ordinary places, to locations refreshingly different from the scenes of daily routines. Iowa's state parks, preserves, recreation and wildlife areas, forests, and waterways are sites of scenic resources and diverse habitats; they also are places where Iowa's geologic past is on view. Many of these sites showcase topographic features and geologic materials that characterize one or another of Iowa's landform regions, and they reflect the spectrum of geologic processes and deposits that have shaped the state. An awareness of these past links to our present environment offers a comfortable sense of familiarity with the land. We see and appreciate the Iowa landscape from a new perspective and realize that the land's geological past is an inseparable part of the state's natural and cultural history. The following annotated list of Iowa's scenic public areas is organized by landform region and invites the interested visitor to explore different chapters of the state's geological history. The map on page 110 locates each of these sites. In addition, many state wildlife management areas as well as county and municipal parks offer excellent opportunities of geological focus. Contact the Iowa Department of Natural Resources, county conservation boards, municipal parks departments, U.S. Department of the Interior, or U.S. Army Corps of Engineers for further information about sites that interest you.

The lush vegetation described by Leonard (left) remains an attraction for today's visitors hiking the steep winding trails of Pikes Peak State Park. This wooded glen, containing Sand Cave, has been carved from a thick, colorful Ordovician sandstone (St. Peter Formation).

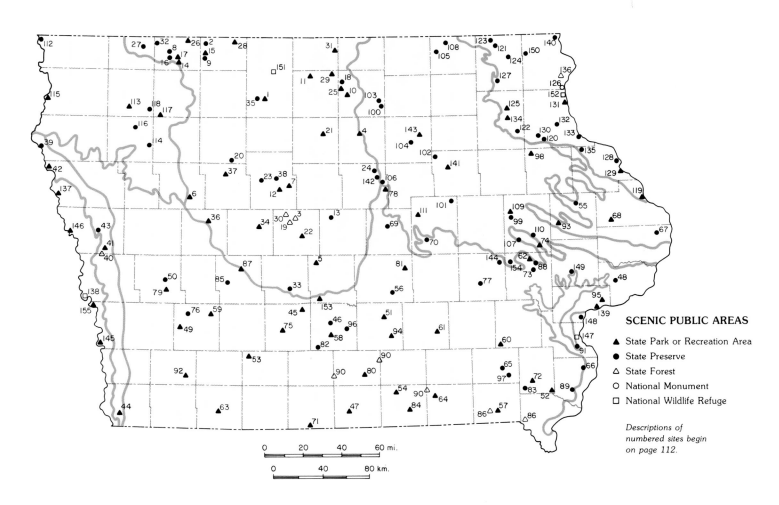

Scenic Public Areas to Visit in Iowa

SCENIC PUBLIC AREAS

▲ State Park or Recreation Area

● State Preserve

△ State Forest

○ National Monument

□ National Wildlife Refuge

Descriptions of numbered sites begin on page 112.

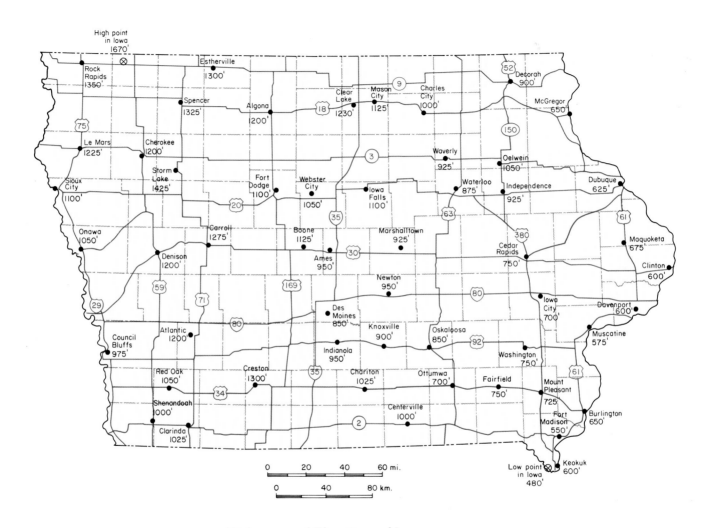

Highways and Elevations of Iowa

111

Des Moines Lobe

Ambrose A. Call State Park, Kossuth County. Steep valley margins along the East Fork Des Moines River are carved from deposits of the Algona end moraine. 1

Anderson Prairie State Preserve, Emmet County. Native prairie species occupy irregular gravelly knobs and kettle depressions characteristic of the Altamont end moraine. The preserve overlooks the West Fork Des Moines River, marking the western edge of the Algona ice advance. 2

Barkley State Forest, Boone County. Elkhorn Creek cuts deeply into upland deposits of the Altamont end moraine. 3

Beeds Lake State Park, Franklin County. The ponded waters of Spring Creek along the eastern margin of the Des Moines Lobe occupy a former glacial meltwater channel. 4

Big Creek Lake State Park, Polk County. This artificial lake within the recently glaciated uplands of the Des Moines Lobe drains into the Saylorville Reservoir/Des Moines River system through an artificial channel. 5

Black Hawk Lake State Park, Sac County. Hummocky terrain of the Bemis end moraine nearly surrounds this glacial lake, part of a large outwash channel complex along the western margin of the Des Moines Lobe. 6

Brushy Creek State Recreation Area and Preserve, Webster County. This deep, narrow valley, containing a prominent series of terraces and benches, developed entirely during melting of the Des Moines Lobe ice. 7

Cayler Prairie State Preserve, Dickinson County. Prominent gravel ridges and wetland swales dominate the sweeping landscape of this native prairie, land once in contact with stagnant glacial ice and typical of deposits of the Altamont end moraine. 8

Cheever Lake State Preserve, Emmet County. Interesting plants and abundant waterfowl inhabit this pothole marsh, part of a large outwash channel network through knob-and-kettle terrain typical of the Altamont end moraine. 9

Clear Lake State Park, Cerro Gordo County. Shallow waters occupy this large glacial lake situated on the Bemis end moraine along the eastern edge of the Des Moines Lobe. 10

Crystal Lake State Park, Hancock County. This kettle lake in a glacial outwash channel is surrounded by hummocky, high-relief topography comprising the eastern lateral margin of the Algona end moraine. 11

Most of Iowa's natural lakes are found on the Des Moines Lobe, an area little changed in topographic appearance since glacial ice melted from the area about 12,500 years ago. West Okoboji Lake, with Arnolds Park in the background, is the deepest of Iowa's glacial lakes.

Timothy J. Kemmis

Dolliver Memorial State Park, Webster County. Bluffs of Pennsylvanian sandstone line the deeply carved valley of Prairie Creek near its junction with the Des Moines River. The valley was excavated rapidly by glacial meltwater from the Des Moines Lobe. 12

Doolittle Prairie State Preserve, Story County. This complex of prairie potholes formed as the Altamont ice advance stagnated and melted, leaving subtly linked depressions across the land surface (see photo, p. 41). 13

Emerson Bay State Park, Dickinson County. The Altamont end moraine surrounds this embayment along the southwest shoreline of West Okoboji Lake. 14

Fort Defiance State Park, Emmet County. The valley of School Creek, a tributary of the Des Moines River, has cut deeply into knob-and-kettle terrain of the Altamont end moraine. 15

Freda Haffner Kettlehole State Preserve, Dickinson

113

County. This steep-sided, bowl-shaped depression on the Altamont end moraine was created when an isolated, perhaps partially buried pocket of clean ice melted slowly in place after the main Des Moines Lobe ice sheet wasted away (see photo, p. 45). 16

Gull Point State Park, Dickinson County. Numerous glacial boulders are present along the shoreline of this peninsula into West Okoboji Lake, the deepest of Iowa's glacial lakes. 17

Hoffman Prairie State Preserve, Cerro Gordo County. This swale, inhabited by prairie and wetland species, and the surrounding knobby hills are typical of the eastern lateral margins of the Altamont end moraine. 18

Holst State Forest, Boone County. Small tributaries of the nearby Des Moines River carved the broken, dissected terrain which dominates this woodland. 19

Kalsow Prairie State Preserve, Pocahontas County. The low relief of this native prairie and its surroundings resulted from glacial action during the Altamont ice advance. This preserve is nearly centered over the buried Manson Crater, formed by meteorite impact in preglacial time. 20

Lake Cornelia State Park, Wright County. A series of prominent ridges, knobby hills, and closed swales resulting from stagnation of the Altamont ice advance surrounds this glacial lake west of the Iowa River valley. 21

Ledges State Park, Boone County. Outcrops of Pennsylvanian sandstone line the walls of Peas Creek valley near its junction with the Des Moines River valley. These deep valleys were carved rapidly by swift meltwaters from the Des Moines Lobe glacier. 22

Liska-Stanek Prairie State Preserve, Webster County. A remnant of native prairie grows across this landscape, typical of the low-relief terrain in the central part of the Des Moines Lobe. 23

Mann Wilderness State Preserve, Hardin County. Short, deep tributary valleys of the Iowa River valley are carved into upland deposits of the Bemis end moraine, forming scenic terrain and diverse woodland habitats. 24

McIntosh Woods State Park, Cerro Gordo County. Irregular terrain of the Bemis end moraine forms the northwestern shoreline of Clear Lake, a glacial lake on the eastern margin of the Des Moines Lobe. 25

Mini-Wakan State Park, Dickinson County. Lowland landscapes of recently glaciated terrain rim the northern shore of Spirit Lake, Iowa's largest natural body of water. 26

Ocheyedan Mound State Preserve, Osceola County. This roughly conical knob is a glacial kame, where sand and gravel filled a large cavity once present within stagnant glacial ice. The preserve is situated along the westernmost margins of the Bemis end moraine in Iowa (see photo, p. 42). 27

Okamanpedan State Park, Emmet County. Lowland landscapes mark the outlet of Okamanpedan Lake, a glacial lake forming the headwaters of the East Fork Des Moines River. 28

Pilot Knob State Park and Preserve, Hancock County. Irregular hummocks and kettle depressions dominate this segment of the Altamont end moraine, including Pilot Knob, a glacial kame, and Deadman's Lake, an unusual sphagnum bog. 29

Pilot Mound State Forest, Boone County. This conical wooded prominence, an excellent example of a glacial kame, is one of the numerous knobby hills which form the Altamont end moraine. 30

Rice Lake State Park, Winnebago County. Beaver Creek originates in this elongated glacial lake and wetland complex situated amid the hummocky terrain along the eastern lateral margins of the Altamont end moraine. 31

Silver Lake Fen State Preserve, Dickinson County. This water-saturated peat deposit accumulated around an unusual upwelling of groundwater and provides a unique botanical habitat above the southwestern shoreline of Silver Lake, a kettle lake within the Bemis end moraine. 32

Silvers-Smith Woods State Preserve, Dallas County.

The eroded valley margin of the North Raccoon River is carved into Des Moines Lobe deposits belonging to the Bemis ice advance and forms the setting for this woodland preserve. 33

Spring Lake State Park, Greene County. This lowland is part of a broad plain of glacial outwash formed just beyond the prominent ridge-front of the Altamont end moraine to the north. The lake occupies a former gravel pit. 34

Stinson Prairie State Preserve, Kossuth County. The prominent south-facing slope of this native prairie marks the edge of the Algona end moraine, the last stand of glacial ice in north-central Iowa, and overlooks a broad outwash plain to the south. 35

Swan Lake State Park, Carroll County. The southernmost of Iowa's large glacial lakes, the shallow waters of Swan Lake occupy the outermost edge of the Des Moines Lobe in an outwash drainageway to the nearby Middle Raccoon River. 36

Twin Lakes State Park, Calhoun County. These elongated glacial lakes formed on low-relief landscapes during melting of the Altamont advance. 37

Woodman Hollow State Preserve, Webster County. Bedrock deposits of Pennsylvanian sandstone are exposed beneath Des Moines Lobe glacial drift. The steep, narrow, wooded ravine opens into the Des Moines River

Donald Poggensee

Some of the most scenic vistas in Iowa are viewed from the summits of loess bluffs overlooking the broad Missouri River valley. Alternating peaks and saddles along ridge crests are tinged with the rust tones of dried prairie grasses during the fall and winter seasons.

valley and hosts a variety of plant communities. 38

Loess Hills

Five-Ridge Prairie State Preserve, Plymouth County. The narrow ridges and deep ravines in the northern end of the Loess Hills region were sculpted from deposits of windblown silt swept from the Missouri and Big Sioux valleys during glacial time (*see* photo, p. 8). Cretaceous bedrock also outcrops in the vicinity. 39

Loess Hills Pioneer State Forest, Harrison-Monona

counties. Originally native prairie with some scattered oaks, these woodland tracts in the heart of the deep-loess landscapes developed in response to the suppression of fire after Euro-American settlement of the area began about 1850. 40

Preparation Canyon State Park, Monona County. Steep timbered slopes and narrow crooked ridges dominate this landscape carved from deep deposits of loess blown from the Missouri Valley to the west. 41

Stone State Park and Mount Talbot State Preserve, Woodbury County. Steep prairie-covered slopes alternating with oak-dominated ravines, typical landscapes of the Loess Hills, provide scenic overlooks of the adjacent Missouri and Big Sioux valleys. Outcrops of Cretaceous limestone, shale, and sandstone are seen in the area. 42

Turin Loess Hills State Preserve, Monona County. Outstanding vistas of the broad Missouri Valley and its abrupt boundary with the steeply pitched terrain carved from thick deposits of loess highlight this preserve. The area is part of a National Natural Landmark. 43

Waubonsie State Park, Fremont County. Corrugated landscapes, a dense drainage network, scenic overlooks of the Missouri River floodplain, and native prairie openings among oak-dominated woodlands occupy this southern segment of the Loess Hills. Pennsylvanian limestone and shale are exposed in the western ravines. 44

Southern Iowa Drift Plain

Badger Creek State Recreation Area, Madison County. The impounded waters of Badger Creek form an artificial lake with numerous shoreline inlets reflecting the steep, well-drained slopes of this Pre-Illinoian drift plain. 45

Berry Woods State Preserve, Warren County. This woodland, dominated by white oaks and hickories, is situated in steeply carved bluffs overlooking the Middle River valley. 46

Bobwhite State Park, Wayne County. Impounded waters of the South Fork Chariton River partially fill this steeply sloping valley landscape carved from Pre-Illinoian glacial drift. 47

Cameron Timber State Preserve, Scott County. The ground beneath this stand of native hardwoods slopes northward toward the valley of Hickory Creek amid rolling landscapes of loess-mantled Illinoian drift. 48

Cold Springs (Crystal Lake) State Park, Cass County. This ponded wetland collects groundwater seeping from the base of Cretaceous sandstone bluffs forming the south wall of the East Nishnabotna Valley and occupies a portion of the river's floodplain. 49

Dineson Prairie State Preserve, Shelby County. This upland remnant of loess-mantled Pre-Illinoian drift is creased with eroded slopes draining eastward into the broad valley of the West Nishnabotna River. 50

Elk Rock State Park, Marion County. Steeply rolling landscapes carved from Pre-Illinoian glacial drift contain the waters of Red Rock Reservoir along the Des Moines River valley. 51

Geode State Park, Henry County. The waters of Cedar Creek, a tributary of the Skunk River, are impounded to form Geode Lake within steep wooded landscapes of dissected Illinoian glacial drift. Mississippian outcrops in the area contain the famous crystal-lined geodes, Iowa's state rock (photo, right). 52

Green Valley State Park, Union County. Green Valley Lake, a reservoir nestled below loess-mantled summits of Pre-Illinoian drift, is the focus of this park near the headwaters of the Platte River. 53

Honey Creek State Park, Appanoose County. The waters of Rathbun Reservoir extend up the valley of Honey Creek, one of the numerous streams that branch across this well-drained region. 54

Indian Bluffs Primitive Area, Jones County. Resistant Silurian dolomite confines the deep, narrow valley of the South Fork Maquoketa River and contributes to the rugged topography near the northeastern limits of the Southern Iowa Drift Plain. 55

Kish-Ke-Kosh Prairie State Preserve, Jasper County.

An abrupt narrowing of the Skunk River floodplain along the east valley wall expresses the resistance of Pennsylvanian bedrock. This prominent projection into the valley is mantled with windblown sand and silt which provide an unusual habitat for native prairie species. 56

Lacey-Keosauqua State Park, Van Buren County. The relief of this steeply rolling woodland is enhanced by its position along the south side of the great bend in the Des Moines River, where Mississippian limestone is exposed. 57

Lake Ahquabi State Park, Warren County. Steeply rolling slopes typical of eroded, loess-mantled Pre-Illinoian glacial plains surround this artificial lake impounded in a side valley of Squaw Creek. 58

Lake Anita State Park, Cass County. This impounded tributary of Turkey Creek forms a lake within the rolling, well-drained slopes of Pre-Illinoian glacial drift. 59

Lake Darling State Park, Washington County. The impounded stream flow of Honey Creek, a tributary of the Skunk River, forms this lake among numerous sloping hillsides typical of the eroded, loess-mantled Pre-Illinoian drift plain. 60

Lake Keomah State Park, Mahaska County. This woodland-rimmed lake, just north of uneroded Pre-Illinoian summits, is impounded in steeply rolling terrain carved by a tributary of the nearby South Skunk River.

Pennsylvanian rocks in the vicinity have been mined for coal. 61

Lake Macbride State Park, Johnson County. Waters forming the two arms of this lake, impounded within rolling valley landscapes carved by Jordan and Mill creeks, flow over a spillway of Devonian limestone into the Coralville Reservoir to the west. 62

Lake of Three Fires State Park, Taylor County. Timbered slopes overlook the impounded waters of a tributary of the East Fork One Hundred and Two River in steeply rolling, loess-mantled landscapes carved from Pre-Illinoian glacial drift. 63

Lake Wapello State Park, Davis County. The waters of Pee Dee Creek, a tributary of Soap Creek to the north, form this artificial lake in steeply rolling landscapes typical of the dissected Pre-Illinoian drift plain. 64

Lamson Woods State Preserve, Jefferson County. This woodland on the outskirts of Fairfield occupies the slopes of a small side-valley of Crow Creek and is cut into Pre-Illinoian glacial deposits. 65

Malchow Mounds State Preserve, Des Moines County. Woodland-period Indians occupied this loess-mantled bluff of Illinoian drift overlooking the broad floodplain of the Mississippi River valley. Mississippian limestone outcrops at the base. 66

Timothy J. Kemmis

Geodes are roughly spherical rocks with hollow, crystal-lined interiors and are highly prized by collectors. This 17-inch-diameter specimen, lined with sparkling crystals of pink and gray quartz, is from an area near Keokuk in southeastern Iowa.

Manikowski Prairie State Preserve, Clinton County. This native prairie is noted for low outcroppings of Silurian dolomite which mark the eastern edge of an ancestral valley of the Mississippi River known as Goose Lake Channel. 67

Maquoketa Caves State Park, Jackson County. An inside view of landscapes formed by karst processes in Silurian dolomite reveals caves, sinkholes, and a natural bridge along this gorge-like valley of Raccoon Creek near the northeastern limits of the Southern Iowa Drift Plain. 68

Marietta Sand Prairie State Preserve, Marshall County. The upland sand deposits which host the native flora were blown by the wind from the nearby Iowa River valley, which carried abundant outwash from melting ice on the Des Moines Lobe to the west. 69

Mericle Woods State Preserve, Tama County. This sloping, well-drained woodland habitat is situated along the dissected valley margins of Deer Creek near the northern margin of the Southern Iowa Drift Plain. 70

Nine Eagles State Park, Decatur County. Steeply rolling landscapes of loess-mantled Pre-Illinoian drift characterize this tributary watershed along the eastern margin of the Thompson River valley near the Missouri border. 71

Oakland Mills State Park, Henry County. Steep wooded bluffs carved from Pre-Illinoian glacial drift overlying Mississippian limestone overlook a narrow segment of the Skunk River valley. 72

Old State Quarry State Preserve, Johnson County. During the nineteenth century, Devonian limestone from this wooded bluff along the edge of the Iowa River valley was quarried for use in building the Old Capitol in Iowa City. The geologic type-section for the State Quarry Limestone is here. 73

Palisades-Kepler State Park and Palisades-Dows State Preserve, Linn County. Steep, densely wooded hillslopes above both sides of the Cedar River valley drop along sheer cliffs of Silurian dolomite which comprise an ancient tropical marine reef. 74

Pammel State Park, Madison County. Following an abruptly changing course, the Middle River has eroded deeply through Pre-Illinoian glacial deposits and exposed the underlying Pennsylvanian limestones and shales. 75

Pellett Woods State Preserve, Cass County. This woodland overlooks the East Nishnabotna River valley. A narrow divide of loess-mantled Pre-Illinoian drift separates the valley from that of Troublesome Creek to the south. 76

Pilot Grove State Preserve, Iowa County. This historic woodland landmark occupies a high, narrow divide of loess-mantled Pre-Illinoian drift separating streams that drain north to Old Man's Creek from those draining south to the North English River. 77

Pine Lake State Park, Hardin County. Pine Creek, a westward-flowing tributary of the nearby Iowa River, has eroded through Pre-Illinoian drift to expose Pennsylvanian sandstone in the steep slopes above its impounded waters. 78

Prairie Rose State Park, Shelby County. A dam holds the waters of this artificial lake within rolling loess-mantled Pre-Illinoian drift landscapes just east of the East Branch West Nishnabotna River valley. 79

Red Haw State Park, Lucas County. Red Haw Lake is the highest in a series of impoundments near the headwaters of Little White Breast Creek in steeply rolling loess-mantled landscapes of Pre-Illinoian glacial drift. 80

Rock Creek State Park, Jasper County. The impounded waters of Rock Creek Lake are confined by steep slopes typical of the dissected, loess-mantled Pre-Illinoian drift plain. 81

Rolling Thunder Prairie State Preserve, Warren County. Steeply rolling well-drained landscapes typical of the loess-mantled Pre-Illinoian drift plain in southern Iowa are host to this tallgrass prairie remnant. 82

Savage Memorial Woods State Preserve, Henry County. Oak and hickory dominate this native woodland along the steeply sloping western margin of the Cedar Creek valley. 83

Sharon Bluffs State Park, Appanoose County. This woodland encompasses uplands, steep sideslopes, and floodplain along the west side of the Chariton River valley. 84

Sheeder Prairie State Preserve, Guthrie County. The rolling slopes of this native prairie are formed of loess-mantled Pre-Illinoian glacial drift and overlook the valley of Lone Grove Creek to the west. 85

Shimek State Forest, Lee and Van Buren counties. Named for one of Iowa's foremost naturalists, the forest includes rolling timbered tracts covering uplands and sideslopes along margins of the Des Moines River valley. Pennsylvanian coals were once mined here. 86

Springbrook State Park, Guthrie County. These steeply rolling Pre-Illinoian landscapes, just off the southwest edge of the Des Moines Lobe, drain to the Middle Raccoon River and expose outcrops of iron-stained Cretaceous sandstone. 87

Stainbrook Geological Preserve, Johnson County. Devonian limestones along the east edge of the Coralville Reservoir display grooves of Pre-Illinoian glaciers and contain diverse fossils of marine organisms that once inhabited warm, shallow seas. 88

Starrs Cave State Preserve, Des Moines County. The

Carbonized plant fossils are particularly abundant in the coal-bearing strata of central and southern Iowa. These 305-million-year-old seed-fern leaves from Dallas County grew in tropical coastal swamps during Pennsylvanian time.

Steven A. Hall and Rainer P. Hanson

deep, narrow, picturesque valley of Flint Creek contains the geologic type-section of the Starrs Cave Formation, a fossiliferous Mississippian limestone named for the cave passage along the northern valley wall (see lithograph, p. 3). 89

Stephens State Forest, Clarke, Lucas, Monroe, Appanoose, and Davis counties. Scattered forested tracts of steeply rolling terrain follow the valleys of White Breast, North Cedar, and Soap creeks which are carved into loess-mantled Pre-Illinoian glacial drift. Pennsylvanian coals were once mined in the area. 90

Toolesboro Mounds State Preserve, Louisa County. Ceremonial Indian mounds (Woodland period) line a high, narrow divide of loess-mantled Illinoian drift above the confluence of the Mississippi and Iowa river valleys. 91

Viking Lake State Park, Montgomery County. The waters of Dunns Creek are impounded within steeply rolling landscapes typical of the dissected, loess-mantled Pre-Illinoian drift plain. 92

Wapsipinicon State Park, Jones County. Blufflands

along narrow valleys of the Wapsipinicon River and Dutch Creek reveal Silurian dolomite beneath loess-mantled Pre-Illinoian glacial drift. 93

Wilcox Park, Marion County. Churned landscapes of abandoned strip mines along the Honey Creek valley contain exposures of coal and some Pennsylvanian plant fossils (photo, left). 94

Wildcat Den State Park, Muscatine County. Pine Creek has carved its valley through Illinoian glacial drift and into Pennsylvanian sandstone en route to the entrenched gorge of the Mississippi Valley. Some Devonian limestone and shale occur at the base of bluffs. 95

Woodland Mounds State Preserve, Warren County. Steep timbered ridges of loess-mantled Pre-Illinoian glacial drift overlook the South River valley to the northwest. 96

Woodthrush Woods State Preserve, Jefferson County. This rolling woodland spans a small tributary valley of Wolf Creek, part of the well-integrated drainage network typical of this region. 97

Iowan Surface

Backbone State Park, Delaware County. Woods, springs, and picturesque promontories of weathered Silurian dolomite along the deep, narrow valley of the Maquoketa River provide a sharp contrast to the gently roll-ing upland landscapes (see photo, p. 5). 98

Behrens Ponds and Woodland State Preserve, Linn County. Shallow wetlands surrounded by open oak-hickory woods and grasslands provide diverse habitats among irregular deposits of windblown sand that originated from the Cedar River valley immediately to the west. 99

Bird Hill State Preserve, Cerro Gordo County. Abundant marine fossils weather from limey Devonian shales (Lime Creek Formation) in the northern part of the Iowan Surface. The overlying Pre-Illinoian glacial deposits are thin, and the influence of bedrock is apparent as a low cuesta in the landscape. 100

Casey's Paha State Preserve, Tama County. This oblong, isolated ridge, oriented northwest to southeast and mantled with windblown loess and sand, rises above the surrounding plain as an erosional remnant of once-thicker Pre-Illinoian glacial deposits. 101

Cedar Hills Sand Prairie State Preserve, Black Hawk County. Wetland depressions disbursed among deposits of windblown sand account for diverse prairie habitats on this narrow upland divide between the broad valleys of West Fork Cedar River and Beaver Creek. 102

Claybanks Forest State Preserve, Cerro Gordo County. Historically known as Hackberry Grove, these bluffs along the south side of the Winnebago River (originally called Lime Creek) contain diverse marine fossils of Devonian age and the type-section of the Lime Creek Formation. 103

Clay Prairie State Preserve, Butler County. A diverse native flora highlights this prairie-covered slope not far from the broad West Fork Cedar River and typical of the gently rolling landscapes of this region. 104

Crossman Prairie State Preserve, Howard County. Seeps on this gentle slope provide a moist habitat for tallgrass prairie species. Glacial erratics here are also characteristic of Iowan Surface landscapes. 105

Fallen Rock State Preserve, Hardin County. The Iowa River valley marks the western edge of the Iowan Surface and the eastern border of the Des Moines Lobe. Pennsylvanian sandstone outcrops along the steep wooded bluffs beneath deposits of Pre-Illinoian glacial drift. 106

Hanging Bog State Preserve, Linn County. Located at the brink of the Cedar River valley, a steep wooded slope surrounds a boggy opening of groundwater-saturated deposits hosting a large population of skunk cabbage. 107

Hayden Prairie State Preserve, Howard County. One of Iowa's largest remnants of native prairie contains occasional glacial erratics and is situated on open, gently rolling slopes typical of the Iowan Surface. 108

Pleasant Creek State Park, Linn County. The ponded waters of Pleasant Creek form a lake within Iowan Surface landscapes which steepen near the western edge of the Cedar River valley. 109

Rock Island State Preserve, Linn County. The proximity of the Cedar River valley to the west accounts for the deposits of windblown sand which underlie this diverse upland prairie. 110

Union Grove State Park, Tama County. The Iowan Surface steepens in the vicinity of Deer Creek, which contains an artificial lake within its valley. Mississippian limestone occurs at shallow depths and is quarried nearby. 111

Northwest Iowa Plains

Gitchie Manitou State Preserve, Lyon County. The oldest bedrock to be seen on the Iowa landscape is the 1.6-billion-year-old pink Sioux Quartzite (see photo, p. 81). The durable quartz-rich outcrops along the Big Sioux River valley were designated as the type-section of this formation in 1870. 112

Mill Creek State Park, O'Brien County. Cole Creek is impounded near its junction with Mill Creek to form a lake within open, rolling landscapes on the western edge of Wisconsinan drift (Sheldon Creek Formation). This segment of Mill Creek was an ice-marginal stream about 25,000 years ago. 113

Nestor Stiles Prairie State Preserve, Cherokee County. This native prairie tract at the western edge of Wisconsinan glacial drift (Sheldon Creek Formation) occupies a lowland sag near the headwaters of a tributary of the Maple River, once an ice-marginal stream. 114

Oak Grove State Park, Sioux County. Steep loess-mantled blufflands of Pre-Illinoian glacial drift overlying soft Cretaceous shale overlook the broad valley of the Big Sioux River. 115

Steele Prairie State Preserve, Cherokee County. This large prairie remnant with a diverse flora occupies an open, rolling, well-drained landscape typical of the region. Thin loess mantles Wisconsinan glacial drift (Sheldon Creek Formation) within a mile of this drift sheet's western border. 116

Wanata State Park, Clay and Buena Vista counties. The narrow, steep-sided valley of the Little Sioux River formed quickly as ice dams gave way, releasing the waters of glacial Lake Spencer to carve into older Wisconsinan deposits of the Sheldon Creek Formation (see map, p. 6). 117

Wittrock Indian Village State Preserve, O'Brien County. Central Plains Indians of the Mill Creek Culture occupied this site within the shelter of Waterman Creek valley between 600 and 1,000 years ago. The valley is carved from Wisconsinan glacial drift (Sheldon Creek Formation). 118

Paleozoic Plateau

Bellevue State Park, Jackson County. Impressive views of the Mississippi Valley are available from overlooks 250 feet above river level. The steep bluffs are composed of durable Silurian dolomite with some Ordovician shale on lower slopes. 119

Bixby State Preserve, Clayton County. The deep, narrow valley of Bear Creek drains the edge of the Silurian Escarpment and intercepts subterranean flows of cold air and groundwater, producing springs and an ice cave. Large slump blocks of Silurian dolomite occur on steep slopes. 120

Bluffton Fir Stand State Preserve, Winneshiek County. High above a tight bend in the Upper Iowa River, a north-facing slope of creviced Ordovician limestone provides the cool, moist conditions that harbor this remnant population of boreal balsam fir. 121

Brush Creek Canyon State Preserve, Fayette County. Nestled into the creviced Silurian Escarpment, this steep wooded gorge of resistant dolomite outcrops and tilted slump blocks provides a variety of habitats, some favoring boreal plant species. 122

Cold Water Spring State Preserve, Winneshiek County. From beneath a towering bluff of Ordovician limestone flows a spring connected, through a series of siphons, to Iowa's largest known cavern system, Cold Water Cave.

Over seven miles of passages have been mapped. 123

Decorah Ice Cave State Preserve, Winneshiek County. A springtime buildup of ice coats the walls and floor of this passage until late summer. The cave follows a vertical fracture through Ordovician limestone exposed along the north side of the Upper Iowa River valley. 124

Echo Valley State Park, Fayette County. The waters of Otter Creek and Glover Creek drop through narrow valleys carved into limestone and dolomite of the Silurian Escarpment. 125

Effigy Mounds National Monument, Allamakee County. Sheer bluffs of Cambrian sandstone and Ordovician dolomite nearly 300 feet high overlook the Yellow River as it enters the gorge of the upper Mississippi River. Indian mounds (Woodland period) shaped as birds and other animals are clustered here. 126

Fort Atkinson State Preserve, Winneshiek County. Limestone outcrops near the Turkey River were quarried to build this historic fort. The quarry was designated in 1905 as the type-section of the Fort Atkinson Limestone, a member of the more widespread Maquoketa Shale of Ordovician age. 127

Little Maquoketa Mounds State Preserve, Dubuque County. This promontory of Ordovician dolomite, hosting an Indian mound group (Woodland period), stands 200 feet above the junction of the Little Maquoketa River

and its abandoned channel, Couler Valley, through which it flowed before its drainage was captured by the Mississippi at Peru Bottoms (see photo, pp. 94–95). 128

Mines of Spain State Recreation Area, Dubuque County. Lead ores were mined in historic and prehistoric times from creviced Ordovician dolomites (Galena Group) among the high-relief landscapes bordering Catfish Creek, the Mississippi Valley, Horseshoe Bluff, and Cattese Hollow. 129

Mossy Glen State Preserve, Clayton County. Steep walls of dolomite frame this picturesque valley carved from the Silurian Escarpment. Sinkholes, a 60-foot dry falls, large slump blocks, and springs are part of the geologic setting. The underlying blue-green Ordovician shale (Maquoketa Formation) also is exposed at creek level. 130

Pikes Peak State Park, Clayton County. One of the highest points along the Mississippi Valley (nearly 500 feet above river level), this site offers an outstanding vista of the Mississippi River and its confluence with the Wisconsin River. Ordovician and Cambrian dolomites, limestones, and sandstones are exposed (photo, p. 109). 131

Retz Memorial Woods State Preserve, Clayton County. An upland forest with diverse woodland wildflowers covers rugged slopes, including massive slump blocks of Ordovician dolomite, above deeply carved meanders of the Turkey River valley. 132

Jean C. Prior

Fish Farm Mounds State Preserve in Allamakee County is located on a sandy alluvial terrace of the Mississippi River. The mound group was constructed of deposits scraped from the surrounding terrace by Archaic and Woodland Indians between 1000 B.C. and A.D. 330.

Turkey River Mounds State Preserve, Clayton County. A long, narrow, forested ridge 200 feet high with nearly perpendicular bluffs of resistant Ordovician dolomite thinly separates the entrenched Turkey and Mississippi valleys near their confluence (photo, p. 131). Archaic period (3,000-year-old) mounds are protected here. 133

Volga River State Recreation Area, Fayette County. The course of the spring-fed Volga River drops through resistant dolomites of the Silurian Escarpment forming steep rocky bluffs along a narrow, deep, and tightly meandered valley containing an artificial lake. 134

White Pine Hollow State Preserve, Dubuque County. The forested Silurian Escarpment, with sinkholes, springs, and massive slump blocks of dolomite, provides cool, moist habitats and high-relief terrain. This site hosts the southernmost population of native white pine and a richly diverse flora and fauna in one of Iowa's largest and most rugged natural areas. 135

Yellow River State Forest, Allamakee County. These tracts of steep forested slopes occupy much of the Paint Creek watershed near its junction with the entrenched valley of the Mississippi River. Ordovician dolomites and Cambrian sandstones influence the landscape shapes. 136

Alluvial Plains

Browns Lake State Park, Woodbury County. This oxbow lake occupies an abandoned channel of the Missouri River. It encloses wetland habitats and an area of hummocky sand deposits and is surrounded by the broad Missouri River floodplain. 137

DeSoto Bend National Wildlife Refuge, Harrison County. DeSoto Bend is a classic example of an abandoned meander loop of the Missouri River, cut off as the river shifted course. The wetlands attract large numbers of migrating waterfowl, and a fine museum houses artifacts recovered from the sunken Civil War–era riverboat *Bertrand.* 138

Fairport State Hatchery, Muscatine County. The floodplain of the Mississippi River is very narrow along this gorge-like segment of its valley. This is one of the youngest segments of the modern valley and was established during Wisconsinan-age glacial meltwater overflow. 139

Fish Farm Mounds State Preserve, Allamakee County. This mound group (photo, p. 127) is located on the rem-

nant of a sandy alluvial terrace of the Mississippi River, sheltered in a small side-valley below high bluffs of Cambrian sandstone. 140

George Wyth Memorial State Park, Black Hawk County. The groundwater table intersects the land surface in the wetlands and former sand pits that characterize the low-relief floodplain along the north side of the Cedar River valley. 141

Hardin City Woodland State Preserve, Hardin County. This unusually sharp bend in the Iowa River reflects a bedrock bench covered with gravels deposited by Des Moines Lobe glacial meltwater. Rich woodlands and diverse wildflowers are found on the uplands, sideslopes, and floodplain. 142

Heery Woods State Park, Butler County. Forest and wetlands intermingle along the Shell Rock River floodplain. The river impinges against an abrupt bluffline of Devonian limestone. 143

Indian Fish Trap, Iowa County. This Native American fish weir in the Iowa River channel is visible only during very low-flow conditions. The funnel-shaped trap and holding pond were built of glacial erratics from drift deposits exposed along the nearby valley bluffs. 144

Lake Manawa State Park, Pottawattamie County. This lake and adjacent lowland occupy a portion of the Missouri River floodplain and are within sight of the Loess

Hills to the east. 145

Lewis and Clark State Park, Monona County. Blue Lake, an oxbow-shaped body of water, occupies a pronounced meander scar marking the course of an earlier migration of the Missouri River channel across its floodplain. 146

Mark Twain National Wildlife Refuge, Louisa County. Wetlands, including Lake Odessa at the base of steep bluffs along the western valley wall, lace the floor of the Mississippi River valley and outline former positions of the braided river channel. 147

Pecan Grove State Preserve, Muscatine County. This is the northernmost stand of reproducing pecans, normally a floodplain species of the southern states. These trees grow along the banks of Muscatine Slough on a broad expanse of Mississippi River floodplain known as Muscatine Island. 148

Rock Creek Island State Preserve, Cedar County. One of the few large stable islands on an interior stream in Iowa, this forested natural area in the Cedar River channel, opposite the mouth of Rock Creek, hosts diverse floodplain tree species and abundant wildlife. 149

Slinde Mounds State Preserve, Allamakee County. The deeply entrenched Upper Iowa River flows in a tight meander loop around this bedrock bench containing Woodland Indian mounds and prairie above on the valley's steep sideslope. Ordovician dolomite is exposed nearby. 150

Union Slough National Wildlife Refuge, Kossuth County. A series of artificially maintained wetland pools occupies a well-defined outwash channel scoured by meltwaters draining the Algona ice advance of the Des Moines Lobe. 151

Upper Mississippi Wildlife Refuge, Clayton County. The main channel of the Mississippi River flows against steep bluffs of the gorge-like valley. Shallower island-laced segments of the braided river channel form valuable habitats for wetland species. 152

Walnut Woods State Park, Polk County. The Raccoon River floodplain contains woodlands, meander scars of earlier channel migrations, and underlying gravels deposited as glacial outwash along the margin of the Des Moines Lobe. 153

Williams Prairie State Preserve, Johnson County. Diverse native prairie species inhabit this low, generally moist meadow within the Iowa River valley, not far from the low hills which form the southern valley margin (see photo, p. 29). 154

Wilson Island State Park, Pottawattamie County. Timbered lowlands of the Missouri River floodplain are bounded by the south end of DeSoto Lake, the adjoining Missouri River, and a connecting drainageway. 155

Epilogue

Spacecraft such as Pioneer 10, Mariner 10, Viking 1, Voyager 2, Galileo, and Magellan are on exploratory missions of the planets. Of the marvelous images returning to Earth, those of the landscapes and individual landforms of the planetary bodies rivet the attention of scientist and casual observer alike. We look at ridges, basins, and channels for signs of wind, water, ice, volcanism, meteor impacts, and dynamic atmospheres.

When the Apollo astronauts landed on the Moon, the world's attention focused not only on the strangeness of a desolate, cratered lunar landscape but on our home planet as a sphere of blue-green oceans, brownish continental masses, and white polar ice caps veiled in swirled patterns of clouds. There probably has been no greater advance in humanity's perception of its environment than during this historic mission. Referring to the myriad forms of life sustained by our lands, waters, and skies, Joseph Wood Krutch thoughtfully observed at the close of his essay, "The Day of the Peepers," "Don't forget, we are all in this together."

So we are—all in this together. These days we are increasingly aware of our natural environment's strengths and vulnerabilities, and we know that we are inseparable from the environment. The land is an integral part of the natural systems at work on our planet. It may register time with a geological clock, but it is no less sensitive to day-by-day events. People may modify the land to suit their purposes, but it is wise to remember that the land must be used in accordance with its capacities as established by geologic history and expressed in landscape shapes and underlying deposits, including groundwater and mineral resources.

Iowa's early explorers and inhabitants had no choice but to consider the land carefully for landmarks, shelter, nourishment, and safety. Today we are free to turn a more inquisitive eye to the landscape—to study its forms, learn of its history, and enjoy its beauty. Yet our view must also be more encompassing, looking broadly and deeply at the land resource with increased understanding of its importance to us and our future.

This forested, pinnacled rib of
Ordovician dolomite, over 200 feet
high, narrowly separates the deeply
entrenched valleys of the meandering
Turkey River (right) from the
Mississippi River (left) near their
junction in Clayton County. This
unique feature is protected as part of
Turkey River Mounds State Preserve.

Photo by Gary Hightshoe

131

Glossary

algific slope: chilled microclimate along steep rocky hillside; produced by cold, moist air seeping through creviced carbonate bedrock flanked with eroded rock rubble (talus); a specialized habitat in northeastern Iowa.

alluvial/alluvium: clay, silt, sand, or gravel deposited by flowing water.

alluvial fan: gently sloping deposits of alluvium spreading outward from the point where a small stream enters a broad valley.

aquifer: water-bearing sediment or rock unit that yields water to wells and springs.

basement rocks: igneous and metamorphic rocks that compose the Earth's crust below the oldest sedimentary deposits.

bedrock: the first solid rock strata encountered at the land surface or beneath younger deposits.

bench: an erosionally cut, terrace-like platform within a valley; overlain by a thin veneer of alluvium.

blowout: shallow basin in sandy terrain caused by wind scour.

boreal: describes northern regions characterized by tundra and coniferous forest biota.

carbonate: calcium carbonate ($CaCO_3$); also called lime.

carbon-14: a radioactive isotope of carbon having a half-life of 5,660 years and useful in dating organic materials ranging to 45,000 years old.

catsteps: a series of slumped steps or treads across steep slopes of loess, especially in western Iowa.

channel sandstone: sandstone bedrock composed of sand originally deposited in a streambed.

colluvium/colluvial: a loose mass of rock or soil material deposited by slope wash and gravity, usually at the base of a steep slope.

concretion: a hard, irregular nodule of mineral matter secondarily accumulated by chemical precipitation around some small nucleus (such as a root); composed of different material from its surroundings.

cross-bedding: layers in a sedimentary deposit that are inclined at an angle to the main stratification; formed by migration of ripples and sand waves along streambeds.

cuesta: an asymmetrical ridge formed along the eroded edge of gently dipping rock strata.

dendritic drainage: drainage patterns in a branching network across the landscape.

diamicton: a descriptive term for poorly sorted sediment.

dissected landscape: a landscape carved by a network of streams, dividing the land into hills and valleys.

dolomite: a variety of limestone rich in magnesium.

drift: deposits of clay, silt, sand, gravel, and boulders left by glaciers or their meltwater streams.

end moraine: a ridge or high-relief terrain composed of glacial drift formed along the margin of a stationary ice front.

entrenched: describes a steep-sided valley, gorge, or gully cut into the surrounding landscape.

eolian: deposited by the wind.

escarpment: an abrupt, steep slope breaking the continuity of the land surface.

esker: a winding, narrow ridge of sand and gravel; formed originally in a meltwater tunnel beneath stagnant glacial ice.

fen: a permanently saturated deposit of peat formed along a hillslope intersecting a slow flow of groundwater.

floodplain: that part of a valley floor adjacent to a river and submerged when the river overflows its channel.

foraminifera: microscopic marine protozoans that live in tiny whorled shells composed of lime.

geomorphology: study of the shapes and origins of the Earth's surface features.

glacial erratics: boulders, cobbles, and pebbles transported by glacial ice and left some distance from their place of origin.

glacial grooves/striations: straight, parallel furrows or scratches ground into bedrock or etched into pebble surfaces by grinding contact with rock fragments during glacial transport.

glacial stages: periods of glacier growth and expansion accompanied by cold climates.

groundwater: water beneath the Earth's surface that fills pore spaces in soil and rock materials.

headward: extension of stream erosion in an upslope direction.

Holocene: approximately the last 10,500 years of geologic time.

hummocky: describes terrain composed of low, irregular hills; applies to glaciated landscapes and windblown sand deposits.

hydrologic cycle: the continuous circulation of the Earth's water from the sea, through the atmosphere, to the land, and eventual return to the sea and atmosphere by rivers and evaporation.

Hypsithermal: period of maximum postglacial warming that culminated about 6,000 years ago.

ice caves: caves in temperate climates where subterranean drainage of cold air and water cause ice to form and remain throughout much of the year.

impound/impoundment: to collect and confine water in an artificial lake or reservoir.

interglacial stages: long periods of warm, mild climate that separate periods of glacial expansion and cold.

joint: a fracture opening, caused by loss of tensile strength, in rock or soil.

kame: a steep-sided hill of sand and gravel formed in an opening in stagnant glacial ice.

karst: describes topography developed from the dissolving of shallow carbonate bedrock by groundwater and characterized by sinkholes, caves, and subterranean drainage.

kettle: a steep-sided, bowl-shaped depression created when an isolated pocket of glacial ice melted slowly in place.

knob-and-kettle topography: an irregular, glaciated landscape of disconnected mounds and poorly drained depressions.

lag deposit: a residual accumulation of pebbles and cobbles that remain after finer deposits have been eroded away.

loess: windblown silt with minor amounts of sand and clay.

loess kindchen: irregular nodules (concretions) of lime (calcium carbonate) occurring in various sizes within loess deposits.

meanders: looping curves and bends in the channel of a river.

outcrop: an exposure of geologic materials.

outlier: isolated remnant of a once-continuous deposit.

outwash: sand and gravel deposited away from a glacier margin by meltwater streams.

oxbow lake: a crescent- or horseshoe-shaped lake which occupies an abandoned river meander on a floodplain.

paha: an elongated, wind-aligned ridge capped with loess and sand.

paleo-landscape: a former land surface buried by younger deposits and identified by outcropping paleosols.

paleosol: a fossil soil originally weathered into deposits on landscape positions that were stable for long periods; usually buried by younger deposits.

Pearlette Family: a series of separate volcanic ash deposits, originating during Pliocene-Pleistocene time from the Yellowstone National Park area of northwestern Wyoming.

pebble band: a concentration of stony debris left as a lag deposit on eroded glacial drift and usually thinly buried by younger deposits.

periglacial: describes cold-climate phenomena, especially frost action, resulting from proximity to an ice sheet or its climatic influences.

pipe: a vertical tunnel or tube-shaped opening, especially in loess deposits, caused by seepage of water below ground.

pipestem: a reddish brown cylindrical concretion of iron which accumulated vertically around plant roots; found in loess deposits.

Pleistocene: an interval of geologic time spanning the "Ice Age" and including multiple glacial and interglacial stages (see stratigraphic key, p. 35).

point-bar: crescent-shaped deposit of alluvium formed on the inside curve of a stream meander.

prairie potholes: wetlands associated with poorly drained depressions and swales in glaciated landscapes.

Pre-Illinoian: the numerous, undifferentiated glacial stages that occurred early in the Pleistocene, prior to Illinoian time (see stratigraphic key, p. 35).

quartzite: a hard, slightly altered sandstone composed of quartz grains and tightly cemented with silica.

Quaternary: approximately the last 1.6 million years of geologic time.

radiometric dating: determining the age of various geologic materials by measurement of the disintegrating radioactive elements they contain.

relief: differences in elevation; unevenness or irregularity of the land surface.

remote sensing: examination of the Earth's surface

from a distance using camera and film combinations sensitive to different bands of the electromagnetic spectrum.

rock-cored meander: an abandoned stream meander around a steep isolated core of bedrock.

sand-dune topography: irregular mounds and shallow basins formed by the action of wind on loose sand.

scarp: a steep, cliff-like slope.

Silurian Escarpment: the abrupt change in slope marking the leading edge of the outcrop belt of resistant Silurian dolomite across the landscape.

sinkhole: a roughly circular, enclosed depression or opening at the land surface caused by collapse of soil and rock material into cavities in underlying limestone bedrock.

slump blocks: large, angular masses of bedrock that have slipped downslope after separating from nearby outcrops.

speleothems: dripstone deposits, such as stalactites and stalagmites, formed in caverns by infiltrating lime-rich groundwater.

stepped erosion surface: interrupted gradient along a hillslope profile which records a change in erosional intensity during the geologic past.

stone line: see pebble band.

stratified: deposited in layers.

stratigraphy: the layered arrangement of sediments and rock units and their chronological sequence.

swale: a shallow, often elongated, sometimes moist depression on the land surface.

tabular divide: uneroded upland landscape having a broad, level surface; tableland.

talus: a heap of coarse angular rock fragments at the base of a steep bedrock bluff.

terrace: a remnant of a former floodplain constructed during an earlier phase of river history and left elevated within a valley by subsequent downcutting of the river.

till: an unsorted, unstratified mixture of clay, silt, sand, and cobbles deposited directly by glaciers.

topography: the general shapes and arrangement of natural features on the land surface.

truncated spur: the beveled end of a ridge that once projected out into a valley and was later planed off by stream erosion.

type-section/type-locality: the geographic site or locality where a specific sequence of strata was originally described in the literature and is most typically exposed; the designated comparative standard for a named geologic unit.

unconsolidated: describes a geologic deposit whose grains are not cemented together.

underfit stream: a stream that appears too small to have eroded the valley in which it flows.

water table: the top surface of unconfined groundwater; usually beneath the land surface, except at lakes and streams.

Wisconsinan: the last glacial stage of the Pleistocene, following the Sangamon interglacial and preceding the Holocene (see stratigraphic key, p. 35).

Supplementary Reading

Anderson, Wayne I. 1983. *Geology of Iowa: Over Two Billion Years of Change.* Iowa State University Press, Ames.

Cooper, Tom C., ed. 1982. *Iowa's Natural Heritage.* Iowa Natural Heritage Foundation, Des Moines, and Iowa Academy of Science, Cedar Falls.

Hamblin, W. Kenneth. 1985. *The Earth's Dynamic Systems.* 4th ed. Burgess Publishing, Minneapolis.

Harr, Douglas C., Dean M. Roosa, and Jean C. Prior. 1990. *Glacial Landmarks Trail: Iowa's Heritage of Ice.* Iowa Department of Natural Resources and State Preserves Advisory Board, Des Moines.

Horick, Paul J. 1974. *Minerals of Iowa.* Iowa Geological Survey Educational Series 2. Iowa City.

Madson, John. 1985. *Up on the River: An Upper Mississippi Chronicle.* Viking Penguin, New York.

———. 1982. *Where the Sky Began: Land of the Tallgrass Prairie.* Houghton Mifflin, Boston.

Mutel, Cornelia F. 1989. *Fragile Giants: A Natural History of the Loess Hills.* University of Iowa Press, Iowa City.

Prior, Jean C., ed. 1984–1990. *Iowa Geology.* Iowa Department of Natural Resources, Geological Survey Bureau, Iowa City.

Runkel, Sylvan T., and Dean M. Roosa. 1989. *Wildflowers of the Tallgrass Prairie.* Iowa State University Press, Ames.

Sayre, Robert F., ed. 1989. *Take This Exit: Rediscovering the Iowa Landscape.* Iowa State University Press, Ames.

Sullivan, Walter. 1984. *Landprints: On the Magnificent American Landscape.* New York Times Book Co., New York.

Troeger, Jack C. 1983. *From Rift to Drift: Iowa's Story in Stone.* Iowa State University Press, Ames.

Wolf, Robert C. 1983. *Fossils of Iowa: Field Guide to Paleozoic Deposits.* Iowa State University Press, Ames.

Selected References

General

Ager, Derek V. 1981. *The Nature of the Stratigraphical Record.* 2d ed. Halsted Press, New York.

Anderson, Duane. 1981. *Eastern Iowa Prehistory.* Iowa State University Press, Ames.

————. 1975. *Western Iowa Prehistory.* Iowa State University Press, Ames.

Bates, Robert L., and Julia A. Jackson, eds. 1980. *Glossary of Geology.* 2d ed. American Geological Institute, Falls Church, Va.

Bennison, Allan P., and Philip A. Chenoweth. 1984. *Geological Highway Map of the Northern Great Plains Region, North Dakota, Minnesota, South Dakota, Iowa, Nebraska.* American Association of Petroleum Geologists, Tulsa.

Biggs, Donald L., ed. 1987. *North-Central Section of the Geological Society of America Centennial Field Guide.* Geological Society of America, Boulder, Colo.

Iowa Agriculture and Home Economics Experiment Station. 1978. *Iowa Soil Association Map.* Ames.

Iowa State Highway Commission. 1916. *Iowa Lakes and Lakebeds.* State of Iowa, Des Moines.

Krutch, Joseph Wood. 1949. "The Day of the Peepers." In *The Best Nature Writing of Joseph Wood Krutch.* 1969. William Morrow, New York.

Larimer, O. J. 1957. *Drainage Areas of Iowa Streams.* Iowa Highway Research Board Bulletin No. 7.

Oschwald, W. R., F. F. Riecken, R. I. Dideriksen, W. H. Scholtes, and F. W. Schaller. 1965. *Principal Soils of Iowa.* Iowa State University Department of Agronomy Special Report 42, Ames.

Prior, Jean C. 1976. *A Regional Guide to Iowa Landforms.* Iowa Geological Survey Educational Series 3. Iowa City.

Ruhe, Robert V. 1969. *Quaternary Landscapes in Iowa.* Iowa State University Press, Ames.

Outlooks on Iowa Landforms

Anderson, Wayne I. 1989. "Iowa Geology: The Early Years." *Journal of the Iowa Academy of Science,* 96: 81–91.

Andreas, Alfred T. 1875. *Illustrated Historical Atlas of the State of Iowa.* Andreas Atlas, Chicago.

Bartlett, Richard A. 1962. *Great Surveys of the American West.* University of Oklahoma Press, Norman.

Calvin, Samuel. 1898. "Geology of Delaware County." *Iowa Geological Survey Annual Report* 8: 121–192.

————. 1909. "Present Phase of the Pleistocene Problem in Iowa." *Geological Society of America Bulletin* 20: 133–152.

Chamberlin, Thomas C. 1895. "Glacial Phenomena in North America." In *The Great Ice Age,* ed. J. Geikie. 3d ed. D. Appleton, New York.

Hall, James. 1858. *Report of the Geological Survey of*

the State of Iowa. Vol. 1, pt 1: Geology. State of Iowa.

Keyes, Charles. 1913. "Annotated Bibliography of Iowa Geology and Mining." *Iowa Geological Survey Annual Report* 22: 9–908.

Lees, James H. 1926. "Altitudes in Iowa." *Iowa Geological Survey Annual Report* 32: 363–550.

Lewis, Henry. 1967. *The Valley of the Mississippi Illustrated.* Minnesota Historical Society, St. Paul.

McGee, W. J. 1891. "The Pleistocene History of Northeastern Iowa." *Eleventh Annual Report of the Director of the U.S. Geological Survey.* Pt 1: *Geology.* Washington, D.C.

Owen, David Dale. 1852. *Report of a Geological Survey of Wisconsin, Iowa, and Minnesota.* Lippincott, Grambo and Co., Philadelphia.

Prior, Jean C. 1988. "The State Geological Survey of Iowa." In *The State Geological Surveys: A History,* ed. Arthur A. Socolow. American Association of State Geologists Special Volume.

———, and Carolyn F. Milligan. 1985. "The Iowa Landscapes of Orestes St. John." In *Geologists and Ideas: A History of North American Geology,* ed. Ellen T. Drake and William M. Jordan. Geological Society of America Centennial Special Volume 1. Boulder, Colo.

White, Charles A. 1870. *Report on the Geological Survey of the State of Iowa.* 2 vols. Mills and Co., Des Moines.

Whitman, Walt. 1881. "Specimen Days." In *The Complete Prose Works of Walt Whitman,* ed. Thomas B. Harned and Horace L. Traubel. Vol. 1. 1902. G. P. Putnam's Sons, New York.

Geologic Origins of Iowa Landforms

Anderson, Duane C., and Patricia M. Williams. 1974. "Western Iowa Proboscidians." *Proceedings of the Iowa Academy of Science* 81: 185–191.

Baker, Richard G., R. Sanders Rhodes II, Donald P. Schwert, Alan C. Ashworth, Terrence J. Frest, George R. Hallberg, and Jan A. Janssens. 1986. "A Full-Glacial Biota from Southeastern Iowa, USA." *Journal of Quaternary Science* 1: 91–107.

Bunker, Bill J., and George R. Hallberg, eds. 1984. *Underburden-Overburden: An Examination of Paleozoic and Quaternary Strata at the Conklin Quarry near Iowa City.* Geological Society of Iowa Guidebook 41. Iowa City.

Calvin, Samuel. 1909. "Aftonian Mammalian Fauna." *Geological Society of America Bulletin* 20: 341–356.

Chumbley, Craig A., Richard G. Baker, and E. Arthur Bettis III. 1990. "Midwestern Holocene Paleoenvironments Revealed by Floodplain Deposits in Northeastern Iowa." *Science* 249: 272–274.

Clayton, Lee, and Stephen R. Moran. 1982. "Chronology of Late Wisconsinan Glaciation in Middle North America." *Quaternary Science Reviews* 1: 55–82.

Davis, Leo Carson, Ralph E. Eshelman, and Jean C. Prior. 1972. "A Primary Mammoth Site with Associated Fauna in Pottawattamie County, Iowa." *Proceedings of the Iowa Academy of Science* 79: 62–65.

Hallberg, George. 1986. "Pre-Wisconsin Glacial Stratigraphy of the Central Plains Region in Iowa, Nebraska, Kansas, and Missouri." In *Quaternary Glaciations in the Northern Hemisphere*, ed. V. Sibrava, D. Q. Bowen, and G. M. Richmond. Quaternary Science Reviews 5. Pergamon, Elmsford, N.Y.

———, and Timothy J. Kemmis. 1986. "Stratigraphy and Correlation of the Glacial Deposits of the Des Moines and James Lobes and Adjacent Areas in North Dakota, South Dakota, Minnesota, and Iowa." In *Quaternary Glaciations in the Northern Hemisphere*, ed. V. Sibrava, D. Q. Bowen, and G. M. Richmond. Quaternary Science Reviews 5. Pergamon, Elmsford, N.Y.

Hay, Oliver P. 1914. *The Pleistocene Mammals of Iowa*. Iowa Geological Survey Annual Report, Des Moines.

Kay, George F., and Earl T. Apfel. 1929. *The Pre-Illinoian Pleistocene Geology of Iowa*. Iowa Geological Survey Annual Report, Des Moines.

———, and Jack B. Graham. 1943. "The Illinoian and Post-Illinoian Pleistocene Geology of Iowa." *Iowa Geological Survey Annual Report* 38: 1–262.

Norton, William H. 1901. "Geology of Cedar County." *Iowa Geological Survey Annual Report* 11: 281–396.

Schwert, Donald P., and Allan C. Ashworth. 1990. "Ice Age Beetles." *Natural History* 1: 10–14.

Des Moines Lobe

Bettis, E. Arthur, III, and Bernard E. Hoyer. 1986. *Late Wisconsinan and Holocene Landscape Evolution and Alluvial Stratigraphy in the Saylorville Lake Area, Central Des Moines River Valley, Iowa*. Iowa Geological Survey Open File Report 86-1. Iowa City.

———, John Pearson, Mark Edwards, David Gradwohl, Nancy Osborn, Timothy Kemmis, and Deborah Quade. 1988. *Natural History of Ledges State Park and the Des Moines Valley in Boone County*. Iowa 9Natural History Association Guidebook 6 and Geological Society of Iowa Guidebook 48. Iowa City.

———, Timothy J. Kemmis, and Brian J. Witzke, eds. 1985. *After the Great Flood: Exposures in the Emergency Spillway, Saylorville Dam*. Geological Society of Iowa Guidebook 43. Iowa City.

Boulton, Geoffrey S. 1972. "Modern Arctic Glaciers as Depositional Models for Former Ice Sheets." *Quarterly Journal of the Geological Society of London* 128: 361–393.

Buchmiller, Robert, Gary Gaillot, and P. J. Soenksen. 1985. *Water Resources of North-Central Iowa*. Iowa Geological Survey Water Atlas No. 7. Iowa City.

Foster, J. D., and R. C. Palmquist. 1969. "Possible Subglacial Origin for 'Minor Moraine' Topography." *Proceedings of the Iowa Academy of Science* 76: 296–310.

Goldthwait, Richard P., ed. 1975. *Glacial Deposits*. Vol. 21 of *Benchmark Papers in Geology*. Dowden, Hutchinson, and Ross, Stroudsburg, Pa.

Gwynne, Charles S. 1942. "Swell and Swale Pattern of the Mankato Lobe of the Wisconsin Drift Plain in Iowa." *Journal of Geology* 50: 200–208.

Hallberg, George R., E. Arthur Bettis III, Timothy J.

Kemmis, Gerald A. Miller, and Richard G. Baker. 1981. "Unique Quaternary Stratigraphic Sections along Brushy Creek, Webster County, Iowa." Iowa State Historical Department, Division of Historic Preservation, Des Moines.

Kemmis, Timothy J. 1991. "Glacial Landforms, Sedimentology, and Depositional Environments of the Des Moines Lobe, Northern Iowa." Ph.D. diss., University of Iowa, Iowa City.

———. 1981. *Glacial Sedimentation and the Algona Moraine in Iowa.* Geological Society of Iowa Guidebook 36. Iowa City.

———, George R. Hallberg, and Alan J. Lutenegger. 1981. *Depositional Environments of Glacial Sediments and Landforms on the Des Moines Lobe, Iowa.* Iowa Geological Survey Guidebook Series 6. Iowa City.

Lees, James H. 1916. "Physical Features and Geologic History of Des Moines Valley." *Iowa Geological Survey Annual Report* 25: 423–615.

Macbride, Thomas H. 1903. "Geology of Kossuth, Hancock and Winnebago Counties." *Iowa Geological Survey Annual Report* 13: 83–122.

———. 1900. "Geology of Osceola and Dickinson Counties." *Iowa Geological Survey Annual Report* 10: 187–240.

Palmquist, Robert C., and K. Conner. 1978. "Glacial Landforms: Des Moines Lobe Drift Sheet, Iowa." *Annals of the Association of American Geographers* 68: 166–179.

Salisbury, Neil E., and James C. Knox. 1969. *Glacial Landforms of the Big Kettle Locality, Dickinson County, Iowa.* Iowa State Advisory Board for Preserves, Development Series Report 6. Des Moines.

Stewart, Robert A. 1988. "Nature and Origin of Corrugated Ground Moraine of the Des Moines Lobe, Story County, Iowa." *Geomorphology* 1: 111–130.

Van Zant, Kent L. 1979. "Late-Glacial and Postglacial Pollen and Plant Macro-fossils from Lake Okoboji, Northwestern Iowa." *Quaternary Research* 12: 358–380.

Loess Hills

Bettis, E. Arthur, III. 1990. *Holocene Alluvial Stratigraphy and Selected Aspects of the Quaternary History of Western Iowa.* Midwest Friends of the Pleistocene, 37th Field Conference Guidebook. Iowa City.

———, and Dean M. Thompson. 1981. "Holocene Landscape Evolution in Western Iowa—Concepts, Methods, and Implications for Archeology." In "Current Directions in Midwestern Archeology: Selected Papers from the Mankato Conference." *Minnesota Archeological Society Occasional Papers in Minnesota Anthropology* 9: 1–14.

———, Jean C. Prior, George R. Hallberg, and Richard L. Handy. 1986. "Geology of the Loess Hills Region." *Proceedings of the Iowa Academy of Science* 93: 78–85.

Daniels, Raymond B., and Robert H. Jordan. 1966. *Physiographic History and the Soils, Entrenched Stream Systems, and Gullies, Harrison County, Iowa.*

U.S. Department of Agriculture, Soil Conservation Service, Technical Bulletin 1348.

Frest, Terrence J., and Jeffrey R. Dickson. 1986. "Land Snails (Pleistocene–Recent) of the Loess Hills: A Preliminary Survey." *Proceedings of the Iowa Academy of Science* 93: 130–157.

Hallberg, George R. 1979. "Wind-Aligned Drainage in Loess in Iowa." *Proceedings of the Iowa Academy of Science* 86: 4–9.

Handy, Richard L. 1976. "Loess Distribution by Variable Winds." *Geological Society of America Bulletin* 87: 915–927.

Izett, Glen A., and Ray E. Wilcox. 1982. *Map Showing Localities and Inferred Distribution of the Huckleberry Ridge, Mesa Falls, and Lava Creek Ash Beds (Pearlette Family Ash Beds) of Pliocene and Pleistocene Age in the Western United States and Southern Canada.* U.S. Geological Survey Miscellaneous Investigation Series, map I-1325.

Prior, Jean C. 1987. "Loess Hills: A National Natural Landmark." *Iowa Geology* 12: 16–19.

Rhodes, R. Sanders, II, and Holmes A. Semken, Jr. 1986. "Quaternary Biostratigraphy and Paleoecology of Fossil Mammals from the Loess Hills Region of Western Iowa." *Proceedings of the Iowa Academy of Science* 93: 94–130.

Roosa, Dean M., and Darwin D. Koenig. 1990. "Bibliography of the Natural and Cultural History of the Loess Hills of Iowa." *Journal of the Iowa Academy of Science* 97: 18–32.

Salisbury, Neil E., and Ronald Dilamarter. 1969. *An Eolian Site in Monona County, Iowa.* Iowa State Advisory Board for Preserves, Development Series Report 7. Des Moines.

Shimek, Bohumil. 1910. "Geology of Harrison and Monona Counties, Iowa." *Iowa Geological Survey Annual Report* 20: 271–485.

———. 1896. "A Theory of the Loess." *Proceedings of the Iowa Academy of Science* 3: 82–89.

Udden, Johan A. 1903. "Geology of Mills and Fremont Counties." *Iowa Geological Survey Annual Report* 13: 123–183.

Southern Iowa Drift Plain

Bain, H. Foster. 1898. "Geology of Decatur County." *Iowa Geological Survey Annual Report* 8: 255–314.

Baker, Richard G., Donald P. Schwert, E. Arthur Bettis III, Timothy J. Kemmis, Diana G. Horton, and Holmes A. Semken. 1991. "Mid-Wisconsinan Stratigraphy and Paleoenvironments at the St. Charles Site, South-Central Iowa." *Geological Society of America Bulletin* 103: 210–220.

Bettis, E. Arthur, III, and John P. Littke. 1987. *Holocene Alluvial Stratigraphy and Landscape Development in Soap Creek Watershed, Appanoose, Davis, Monroe, and Wapello Counties, Iowa.* Iowa Department of Natural Resources, Geological Survey Bureau Open File Report 87-2. Iowa City.

Boellstorff, John D. 1978. "Proposed Abandonment of Pre-Illinoian Pleistocene Stage Terms." *Geological Society of America Abstracts with Programs* 10: 247.

Cagle, J. W., and A. J. Heinitz. 1978. *Water Resources of South-Central Iowa*. Iowa Geological Survey Water Atlas 5. Iowa City.

Coble, Ronald W. 1971. *The Water Resources of Southeast Iowa*. Iowa Geological Survey Water Atlas 4. Iowa City.

Hallberg, George R. 1980. *Pleistocene Stratigraphy in East-Central Iowa*. Iowa Geological Survey Technical Information Series 10. Iowa City.

———, ed. 1980. *Illinoian and Pre-Illinoian Stratigraphy of Southeast Iowa and Adjacent Illinois*. Iowa Geological Survey Technical Information Series 11. Iowa City.

———, and John D. Boellstorff. 1978. "Stratigraphic 'Confusion' in the Region of the Type Areas of Kansan and Nebraskan Deposits." *Geological Society of America Abstracts with Programs* 10: 255.

———, Thomas E. Fenton, Timothy J. Kemmis, and Gerald A. Miller. 1980. *Yarmouth Revisited*. 27th Field Conference, Midwest Friends of the Pleistocene, Iowa Geological Survey Guidebook Series 3. Iowa City.

Plocher, Orrin W., ed. 1989. *Geologic Reconnaissance of the Coralville Lake Area*. Geological Society of Iowa Guidebook 51. Iowa City.

Ruhe, Robert V., Raymond B. Daniels, and John G. Cady. 1967. *Landscape Evolution and Soil Formation in Southwestern Iowa*. U.S. Department of Agriculture, Soil Conservation Service, Technical Bulletin 1349. Washington, D.C.

Wahl, Kenneth D., G. A. Ludvigson, G. L. Ryan, and W. C. Steinkampf. 1978. *Water Resources of East-Central Iowa*. Iowa Geological Survey Water Atlas 6. Iowa City.

Witzke, Brian J., ed. 1984. *Geology of the University of Iowa Campus Area, Iowa City*. Iowa Geological Survey Guidebook Series 7. Iowa City.

Iowan Surface

Dirks, Richard A., and Carl R. Busch. 1969. "The Giant Boulders of the Iowan Drift and a Consideration of Their Origin." *Proceedings of the Iowa Academy of Science* 76: 282–295.

Hallberg, George R., Thomas E. Fenton, Gerald A. Miller, and Alan J. Lutenegger. 1978. "The Iowan Erosion Surface: An Old Story, an Important Lesson, and Some New Wrinkles." In *42nd Annual Tri-State Geological Field Conference on Geology of East-Central Iowa*, ed. Raymond Anderson, 2-2–2-94. Iowa City.

Prior, Jean C., Dean M. Roosa, Daryl D. Smith, Paul C. Christiansen, and Lawrence Eilers. 1986. *Natural History of the Cedar and Wapsipinicon River Basins on the Iowan Erosion Surface*. Iowa Natural History Association Field Trip Guidebook 4.

Ruhe, Robert V., Wayne P. Dietz, Thomas E. Fenton, and George F. Hall. 1968. *Iowan Drift Problem, Northeastern Iowa*. Iowa Geological Survey Report of Investigations 7. Iowa City.

Northwest Iowa Plains

Anderson, Duane C., and Holmes A. Semken, Jr., eds. 1980. *The Cherokee Excavations: Holocene Ecology and Human Adaptations in Northwestern Iowa.* Academic Press, New York.

Benn, David W., ed. 1990. *Woodland Cultures on the Western Prairies: The Rainbow Site Investigations.* Office of the State Archaeologist Report 18. Iowa City.

Bettis, E. Arthur, III, and Dean M. Thompson. 1982. *Interrelations of Cultural and Fluvial Deposits in Northwest Iowa.* Association of Iowa Archeologists Field Trip Guidebook, University of South Dakota Archeology Laboratory, Vermillion.

Beyer, Samuel W. 1897. "The Sioux Quartzite and Certain Associated Rocks." *Iowa Geological Survey Annual Report* 6: 69–112.

Brenner, Robert L., R. F. Bretz, B. J. Bunker, D. L. Iles, G. A. Ludvigson, R. M. McKay, D. L. Whitley, and B. J. Witzke. 1981. *Cretaceous Stratigraphy and Sedimentation in Northwest Iowa, Northeast Nebraska, and Southeast South Dakota.* Iowa Geological Survey Guidebook Series 4. Iowa City.

Carman, J. Ernest. 1931. "Further Studies on the Pleistocene Geology of Northwestern Iowa." *Iowa Geological Survey Annual Report* 35: 15–193.

Hoyer, Bernard E. 1980. *Geomorphic History of the Little Sioux River Valley.* Geological Society of Iowa Guidebook 34. Iowa City.

Koch, Donald L. 1969. *The Sioux Quartzite Formation in Gitchie Manitou State Preserve.* Iowa State Advisory Board for Preserves, Development Series Report 8. Des Moines.

Macbride, Thomas H. 1902. "Geology of Cherokee and Buena Vista Counties." *Iowa Geological Survey Annual Report* 12: 303–353.

———. 1901. "Geology of Clay and O'Brien Counties." *Iowa Geological Survey Annual Report* 11: 463–508.

Munter, James A., Gregory A. Ludvigson, and Bill J. Bunker. 1983. *Hydrogeology and Stratigraphy of the Dakota Formation in Northwest Iowa.* Iowa Geological Survey Water-Supply Bulletin 13. Iowa City.

Wilder, Frank A. 1900. "Geology of Lyon and Sioux Counties." *Iowa Geological Survey Annual Report* 10: 81–155.

Paleozoic Plateau

Bounk, Michael J., and E. Arthur Bettis III. 1984. "Karst Development in Northeastern Iowa." *Proceedings of the Iowa Academy of Science* 91: 12–15.

Brown, C. E., and J. W. Whitlow. 1960. *Geology of the Dubuque South Quadrangle Iowa-Illinois.* U.S. Geological Survey Bulletin 1123-A. Washington, D.C.

Calvin, Samuel. 1894. "Geology of Allamakee County." *Iowa Geological Survey Annual Report* 4: 35–120.

———. 1904. "Geology of Winneshiek County." *Iowa Geological Survey Annual Report* 16: 37–146.

———, and H. Foster Bain. 1900. "Geology of Dubuque County." *Iowa Geological Survey Annual Report* 10: 379–622.

Delgado, David J., ed. 1983. *Ordovician Galena Group of the Upper Mississippi Valley: Deposition, Diagenesis, and Paleoecology.* Guidebook for the 13th Annual Field Conference of the Great Lakes Section of the Society of Economic Paleontologists and Mineralogists.

Hallberg, George R., E. Arthur Bettis III, and Jean C. Prior. 1984. "Geologic Overview of the Paleozoic Plateau Region of Northeastern Iowa." *Proceedings of the Iowa Academy of Science* 91: 3–11.

Harmon, R. S., H. P. Schwarcz, D. C. Ford, and D. L. Koch. 1979. "An Isotopic Paleotemperature Record for Late Wisconsinan Time in Northeast Iowa." *Geology* 7: 430–433.

Hedges, James A. 1967. "Geology of Dutton's Cave, Fayette County, Iowa." *National Speleological Society Bulletin* 29: 73–90.

Horick, Paul J. 1989. *Water Resources of Northeast Iowa.* Iowa Department of Natural Resources, Geological Survey Bureau Water Atlas 8. Iowa City.

Hoyer, Bernard E., E. Arthur Bettis III, and Brian J. Witzke. 1986. *Water Quality and the Galena Group in the Big Spring Area, Clayton County.* Geological Society of Iowa Guidebook 45. Iowa City.

Hudak, Curtis M. 1987. "Quaternary Landscape Evolution of the Turkey River Valley, Northeastern Iowa." Ph.D. diss., University of Iowa, Iowa City.

Knudson, George E., and James Hedges. 1973. "Decorah Ice Cave State Preserve." *Proceedings of the Iowa Academy of Science* 80: 178–181.

Koch, Donald L. 1973. "Cold Water Cave—Beauty,

Origin, Research." *Iowan Magazine* 22: 23–35.

Leonard, A. G. 1906. "Geology of Clayton County." *Iowa Geological Survey Annual Report* 16: 213–317.

Lively, Richard S. 1983. "Late Quaternary U-Series Speleothem Growth Record from Southeastern Minnesota." *Geology* 11: 259–262.

Roosa, Dean M., Jean C. Prior, Daryl D. Smith, Roger M. Knutson, Dale Henning, and E. Arthur Bettis III. 1983. *Natural History of the Upper Iowa Valley between Decorah and New Albin.* Iowa Natural History Association Field Trip Guidebook 1.

Trowbridge, Arthur C. 1966. *Glacial Drift in the "Driftless Area" of Northeast Iowa.* Iowa Geological Survey Report of Investigations 2. Iowa City.

Alluvial Plains

Anderson, Richard C. 1968. "Drainage Evolution in the Rock Island Area, Western Illinois, and Eastern Iowa." In *The Quaternary of Illinois,* ed. Robert E. Bergstrom. University of Illinois, College of Agriculture Special Publication 14. Urbana, Ill.

Benn, David W., E. Arthur Bettis III, and R. Vogel. 1988. *Archaeology and Geomorphology in Pools 17 and 18, Upper Mississippi River.* U.S. Army Corps of Engineers, Rock Island District, Contract No. DACW25-C-0017, Center for Archaeological Research, Southwest Missouri State University, Springfield.

Bettis, E. Arthur, III, and David W. Benn. 1984. "An Archaeological and Geomorphic Survey in the Central Des Moines River Valley, Iowa." *Plains Anthropologist*

29: 211–227.

———, Richard G. Baker, Brenda K. Nations, and David W. Benn. 1990. "Early Holocene Pecan (*Carya illinoensis*) in the Mississippi River Valley near Muscatine, Iowa." *Quaternary Research* 33: 102–107.

Esling, Steven P. 1983. "*Quaternary Stratigraphy of the Lower Iowa and Cedar River Valleys, Southeast Iowa.*" Ph.D. diss., University of Iowa, Iowa City.

Flock, M. A. 1983. "The Late Wisconsinan Savanna Terrace in Tributaries to the Upper Mississippi River." *Quaternary Research* 20: 165–176.

Hallberg, George R., Jane M. Harbough, and Patricia M. Witinok. 1979. *Changes in the Channel Area of the Missouri River in Iowa, 1879–1976.* Iowa Geological Survey Special Report Series 1. Iowa City.

Hansen, Robert E., and Walter L. Steinhilber. 1977. *Geohydrology of Muscatine Island, Muscatine County, Iowa.* Iowa Geological Survey Water-Supply Bulletin 11. Iowa City.

Leverett, Frank. 1921. "Outline of the Pleistocene History of the Mississippi Valley." *Journal of Geology* 29: 615–626.

Lively, Richard S., coordinator. 1985. *Pleistocene Geology and Evolution of the Upper Mississippi Valley.* Abstracts and Field Trip Guide, Minnesota Geological Survey, St. Paul.

Nations, Brenda K., Richard G. Baker, and E. Arthur Bettis III. 1989. "A Holocene Pollen and Plant Macrofossil Record from the Upper Mississippi Valley." *Current Research in the Pleistocene* 6: 58–59.

Prior, Jean C. 1977. "A Preliminary Geological Survey along the Great River Road in Iowa." In *Archaeology, Geology and Natural Areas*, prepared by John A. Hotopp. Vol. 2 of *Iowa's Great River Road, Cultural and Natural Resources*. Contract Completion Report 108. Office of the State Archaeologist, Iowa City.

Roosa, Dean M., Steven P. Esling, E. Arthur Bettis III, and Jean C. Prior. 1984. *Natural History of the Lake Calvin Basin of Southeast Iowa.* Iowa Natural History Association Field Trip Guidebook 2.

Ruhe, Robert V., and Jean C. Prior. 1970. "Pleistocene Lake Calvin, Eastern Iowa." *Geological Society of America Bulletin* 81: 919–924.

Runkle, Donna L. 1985. *Hydrology of the Alluvial, Buried Channel, Basal Pleistocene and Dakota Aquifers in West-Central Iowa.* U.S. Geological Survey Water Resources Investigations Report 85-4239. Denver, Colo.

Schoewe, Walter H. 1920. "The Origin and History of Extinct Lake Calvin." *Iowa Geological Survey Annual Report* 29: 49–222.

Trowbridge, Arthur C. 1959. "The Mississippi River in Glacial Times." *Palimpsest* 40: 257–288.

Index

Illustrations are indicated by **boldfaced** numbers.

Jean Cutler Prior, a native of Ohio and a graduate of Purdue University and the University of Illinois, has twenty-five years of research, service, and writing experience focusing on Iowa's geology. A senior research geologist for the Geological Survey Bureau of the Iowa Department of Natural Resources, she guided viewers in the Iowa Public Television series "Land between Two Rivers," served on the State Preserves Advisory Board, and edits *Iowa Geology.*

The *Index to Topographic Maps of Iowa,* a list of publications, and ordering information are available upon request from the Geological Survey Bureau, Iowa Department of Natural Resources, 123 N. Capitol Street, Iowa City, Iowa 52242.